World Geography
Explore Your World

AUTHORS: MARK A. STANGE and REBECCA LARATTA
EDITORS: MARY DIETERICH and SARAH M. ANDERSON
PROOFREADER: MARGARET BROWN

ISBN 978-1-62223-533-9

Printing No. CD-404236

Mark Twain Media, Inc., Publishers
Distributed by Carson-Dellosa Publishing LLC

Table of Contents

Introduction...1

United States and Canada...2

Central and South America...18

Europe...33

Russia...48

Southwest Asia and North Africa...63

Africa, South of the Sahara...78

Asia, Oceania, Australia, and Antarctica...93

Maps...108

Global Summit Activity...116

Glossary of Geography Terms...119

Websites and Suggested Activities...121

Answer Keys...122

Introduction

This book is designed to help students better understand the importance of geography and the world in which they live. Geography is important because it encompasses every aspect of our lives. National decisions are often made with geography in mind.

Today, in order for our students to achieve, students must have a global perspective in our ever-changing, ever-growing world. Without a basic knowledge of geography, citizens cannot make informed decisions when participating in the democratic process of our nation.

The study of geography is often divided into five themes. Geographers use these five themes to study the world around us. The themes will be developed throughout this book. They are Location, Place, Human-Environment Interactions, Movement, and Regions. These themes were developed by the Joint Committee on Geographic Education of the National Council for Geographic Education and the Association of American Geographers.

Location: where things are located on Earth, both in absolute and relative terms

Place: the physical and human characteristics that make a place distinct from all others

Human-Environment Interactions: how the land affects the people and how the people affect the land

Movement: how the movement of natural forces, people, goods, and ideas affect a place

Regions: what common physical and human characteristics link a place to other parts of the world

This resource focuses on areas of the world through the lenses of the five themes of geography. Students will be able to make the connections of location, place, interactions, movement, and regions with each specific world region. Using the five themes of geography, teachers can give students the opportunity to think globally and help prepare them for the diverse society in which we live.

Additionally, this resource is aligned to the Common Core State Standards in English Language Arts and Literacy in History/Social Studies, Science and Technical Studies. "While the Standards make references to some particular forms of content, including mythology, foundational U.S. documents, and Shakespeare, they do not—indeed, cannot—enumerate all or even most of the content that students should learn. The Standards must therefore be complemented by a well-developed, content-rich curriculum consistent with the expectations laid out in this document" (from CCSS ELA, Washington, D.C., Common Core State Standards Initiative, pg. 6). Text and activities in this workbook are designed to supplement these efforts. For example, the CCSS ELA Writing Standard W.6.7 "conduct short research projects to anwer a question, drawing on several sources and refocusing the inquiry when appropriate" is reflected on page 6 with the Challenge Activity. Specific Common Core Standards can be found at www.commoncore.org.

United States and Canada: *Location*

The first of the five themes of geography is the study of location. There are two types of location. **Absolute location** is defined as the exact latitude and longitude coordinates on the earth's surface. It answers the question "Where is it?" and gives the exact address of a particular place. **Latitude** lines are the horizontal lines north and south of the equator. **Longitude** lines are the vertical lines measured east and west of the prime meridian. **Relative location** is describing a location in relation to other places. For example, if you describe the location of your home to a friend, you would not need to give the latitude and longitude coordinates. You would instead describe your home's location by identifying streets, buildings, and landmarks that are around your house.

To identify the absolute location of the United States and Canada, you would need to identify the general coordinates on a globe or a map. The United States and Canada span over many lines of latitude and longitude; therefore, it is necessary to determine a specific city or place.

For example, the absolute location of Houston, Texas, is 30 degrees North latitude and 95 degrees West longitude (30°N, 95°W). Quebec City, Quebec, has an absolute location of 47 degrees North latitude and 71 degrees West longitude (47°N, 71°W). The latitude is always written first, followed by the longitude.

The relative location of the United States can be defined as being north of Mexico, south of Canada, bordered by the Atlantic Ocean on its east coast and the Pacific Ocean on its west coast. Canada's relative location is identified as north of the United States, bordered by the Arctic Ocean in the north, the Pacific Ocean on the west, and the Atlantic Ocean on the east.

Name: ______________________________ Date: ____________________

United States and Canada: *Location—Activity*

Directions: Use the map of the United States and Canada, the Internet, or an atlas to answer the questions below.

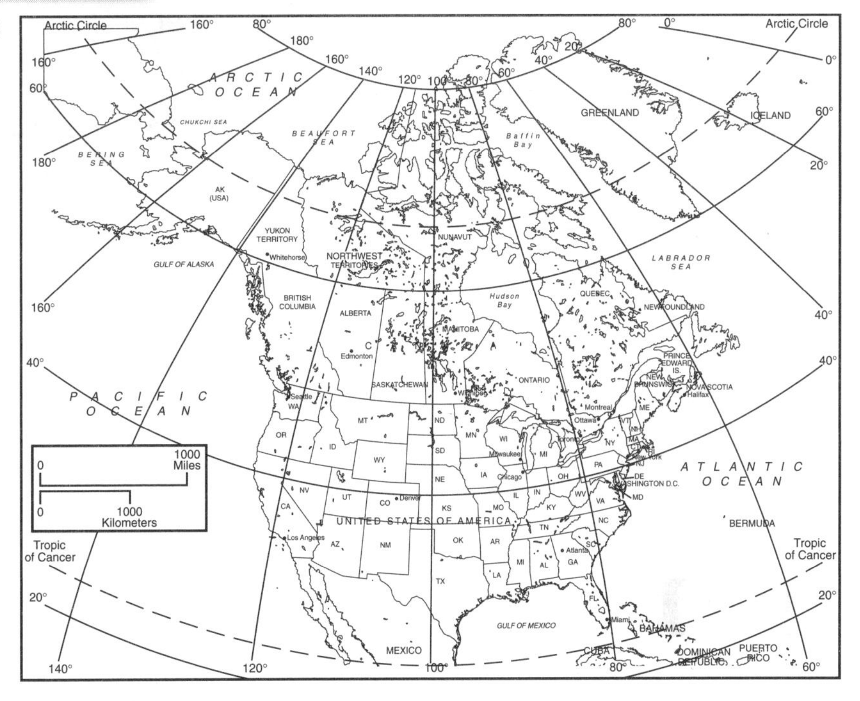

1. Identify the absolute location of the following cities:
   a. Chicago, Illinois 41.8 N, -87.6 W
   b. Washington, D.C. 38, -77 (38N, 77W)
   c. Los Angeles, California 34.05N, 118.W
   d. Denver, Colorado ____________
   e. Atlanta, Georgia ____________
   f. Ottawa, Canada ____________
   g. Winnipeg, Canada ____________
   h. Montreal, Canada ____________

2. Find which cities match the following latitude and longitude coordinates.
   a. 40°N, 74°W 40°, -74° ____________
   b. 47°N, 122°W ____________
   c. 43°N, 88°W ____________
   d. 25°N, 80°W ____________
   e. 53°N, 113°W ____________
   f. 43°N, 79°W ____________
   g. 44°N, 63°W ____________
   h. 60°N, 135°W ____________

Name: ______________________________ Date: ____________________

United States and Canada: *Location—Geoquest*

Directions: Identify the absolute location of your hometown or city. Create a map including a map key and a map title.

Name: ______________________________ Date: ______________________

United States and Canada: *Location—Geoquest (cont.)*

Directions: Identify the relative location of your home in relation to your school. Create a map that includes a map key and a map title.

United States and Canada: *Place*

What makes the United States and Canada unique?

The second theme of geography is **place**. Every place on earth has its own special characteristics that make it unique. To gain an understanding of each place, one must consider its physical characteristics and human characteristics. **Physical characteristics** can be identified as those things that are part of the natural environment. For example, climate, plants, animals, bodies of water, and mountains would be physical characteristics of place.

Human characteristics can be defined as the man-made features or the culture of a place. The architecture of buildings, canals, and bridges; religion; language; and government are examples of human characteristics of place. Both physical and human characteristics provide answers to the question, "What makes this place unique?" If we only look at physical characteristics, you would not have an accurate picture of a specific place. For example, if you identify a place as snowy, cold, and mountainous, you may be thinking of skiing in Vail, Colorado. However, you might be thinking of ice fishing in northern Canada. In order to correctly describe Vail, Colorado, you would have to describe human characteristics such as the English language, a democratic government, Colorado's state capital of Denver, and the local ski resorts.

It is important to note that this theme of place can change over time. Physical characteristics may change due to natural disasters, weathering, and human impact. The Grand Canyon in the western part of the United States or Mount Rushmore in South Dakota has changed over time due to the force of wind, weather, and water eroding away the rock. Furthermore, human characteristics can change as well, depending on factors such as governmental decisions, conflict, and technology. How have the physical and human characteristics of your town changed over the years?

Challenge Activity:

Analyze the impact of these three disasters. How did the physical and human characteristics of each place change as a result of each of the disasters? What lessons were learned as a result?

- Joplin, Missouri, tornado – 2011
- California wildfires – 2014
- Hurricane Katrina – 2005

Name: ______________________________ Date: ______________________

United States and Canada: *Place—Activity*

Places, Places, and More Places

A. Using a map, identify five physical characteristics of the United States and Canada.

	United States	Canada
1.	______________	______________
2.	______________	______________
3.	______________	______________
4.	______________	______________
5.	______________	______________

B. Using an encyclopedia, the Internet, or a textbook, identify five human characteristics of the United States and Canada.

	United States	Canada
6.	______________	______________
7.	______________	______________
8.	______________	______________
9.	______________	______________
10.	______________	______________

C. Identify a physical characteristic (a.) and a human characteristic (b.) of the following places.

11. St. Louis, Missouri
    a. ______________
    b. ______________

12. San Francisco, California
    a. ______________
    b. ______________

13. Rapid City, South Dakota
    a. ______________
    b. ______________

14. Halifax, Nova Scotia
    a. ______________
    b. ______________

15. Ottawa, Ontario
    a. ______________
    b. ______________

Name: ______________________________ Date: ______________

United States and Canada: *Place—Geoquest*

Take the Geoquest Challenge!

1. You are a travel agent. Create a travel brochure that describes a specific place using physical and human characteristics. You may use any variety of traditional and/or digital media technology available to demonstrate your knowledge.
2. Research the following places using the Internet, an atlas, or other resources to see what makes them unique. List them on your own paper.

   a. Yosemite
   b. The White House
   c. The Golden Gate Bridge
   d. Niagara Falls
   e. Carlsbad Caverns
   f. Chateau Frontenac
   g. Hudson Bay
   h. The Citadel
   i. St. Lawrence River
   j. Parliament Hill
   k. Pacific Rim National Park
   l. Grand Canyon
   m. Mount Rushmore
   n. Disney World
   o. Statue of Liberty (see example)

3. Using a red pen or pencil, locate and label the above 15 places on the map. Use the corresponding letter to label the location. (Use a connecting line, if necessary.)

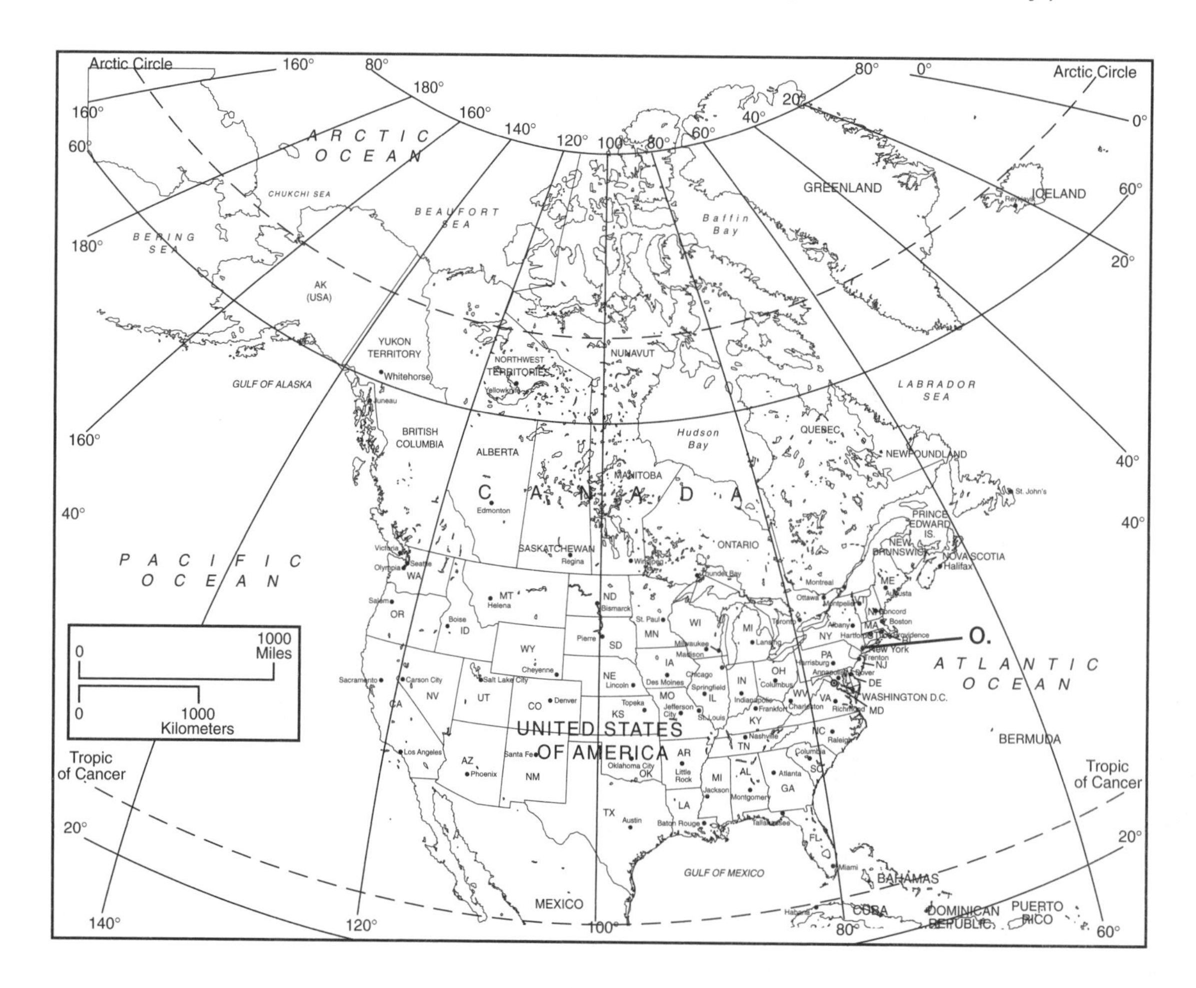

United States and Canada: *Interactions*

The third theme of geography is **human-environment interactions**. This theme deals with how humans affect our environment and how the environment affects human life. For example, humans have cut down forests for the lumber industry in the Pacific Northwest region of the United States. This has resulted in the destruction of animal habitats within the forests. On the other hand, lumber is a much-needed resource for human-kind. Another example of how humans have negatively impacted their environment is the oil spill from the tanker *Exxon-Valdez.* This incident resulted in the contamination of many miles of Alaskan shoreline. Oil coating the ocean water destroyed the habitats of many fish, birds, and other animals. A similar example is the BP oil spill in the Gulf of Mexico in 2010, which contaminated a large part of the Louisiana, Mississippi, Alabama, and Florida shoreline along the Gulf Coast. It was the largest accidental marine oil spill in the history of the petroleum industry.

Humans can also positively affect their environment. One example includes recycling, which allows people to reuse materials instead of wasting valuable resources. Substances like plastic, aluminum, and newspaper can be recycled to be used again. People also create forest preserves and natural wildlife refuges to protect precious animal and plant life.

The environment, or physical surroundings, is an important factor in the lifestyle of people. The climate where you live usually determines the way you live. For example, a Canadian living in the northern Rocky Mountains would be wearing a heavy winter coat, but someone living on a beach in Miami, Florida, may only be wearing a swimming suit. Likewise, because it is so hot and dry, and there is little water in Arizona, fishing would not be a dependable occupation as it is in Nova Scotia.

Challenge Activities:

Activity One: Research the impact of the Keystone Pipeline in the U.S. and Canada. How is this an example of human-environment interactions?

Activity Two: Compare and contrast the *Exxon-Valdez* oil spill and the BP Gulf of Mexico oil spill.

Name: ______________________ Date: ______________________

United States and Canada: *Interactions—Activity*

Directions: Think about your environment. How is your environment different from the following two areas of North America? Fill in examples on the chart below.

	Your City ______________	**New Orleans, Louisiana**	**Vancouver, British Columbia**
Summer Climate			
Winter Climate			
Physical Environment			
Types of Housing			
Foods			
Clothing			
Transportation			
Crops Grown			
Industry/ Products			

Name: ______________________________ Date: ____________________

United States and Canada: *Interactions—Geoquest*

Directions: Answer the following questions using current events examples.

1. In what ways do humans positively impact their environment? ____________________

2. In what ways do humans negatively impact their environment? ____________________

3. In what ways does the environment positively impact human life? ____________________

4. In what ways does the environment negatively impact human life? ____________________

5. Think about your community and the questions above. Explain what type of interaction is currently an issue within your community. Analyze the situation and give your opinion.

United States and Canada: *Movement*

Movement is the fourth theme of geography. It is the study of how people, goods, and ideas have moved or are currently moving around our globe. Concepts such as religion, language, and government spread as people move. As our world becomes more technologically advanced, information can be spread in a matter of seconds. This theme of movement has been a constant factor since the beginning of time. Ancient civilizations broadened their influence through trade, migration, and war. The United States and Canada have become technologically advanced, so they rely on methods such as cell phones and the Internet for movement of information. Most citizens have access to the Internet, which means they can communicate through email, Facebook™, Twitter™, Skype™, and other applications.

Technology has made movement of information almost instantaneous.

Movement not only affects humans but also animal life as well. For example, when an animal habitat is destroyed, those animals are forced to move, and as a result, they impact a new environment.

In order to study this theme of movement, geographers use the questions: who, what, when, where, why, and how do people, things, and ideas move? This information can be used to find patterns of movement, to determine how movement has changed over time, and to determine how it impacts humans.

In the United States, there are clear patterns of movement, which make the lives of Americans stable and orderly. We have a set system of mail delivery, immigration laws, and an extensive road, rail, and air travel system. However, if one of these systems breaks down, movement can be interrupted, and additional problems can occur. For example, if air travel is halted, people and goods cannot get where they are going. This impacts many businesses, which are then unable to produce goods or complete projects. Then, at least temporarily, alternate means of transportation need to be devised.

Movement also plays a part in the physical and human characteristics of a place. Movement can change the human characteristics by bringing in new religions, new foods, new government practices, and even tolerance of cultural diversity. The terrorist attacks on the United States on September 11, 2001, and the resulting reactions are examples of movement in regards to travel, security, religious ideals, and American military response worldwide. Movement also can be impacted by the physical characteristics of a place. For instance, the Rocky Mountains and the Mississippi River presented challenges for early settlers as they tried to move goods from one place to another.

Name: ______________________ Date: ______________________

United States and Canada: *Movement—Activity*

Directions: Using the information from the previous page, answer the following questions.

1. In your own words, define *movement.* ______________________

   ____

2. The French culture has an impact on different parts of North America, such as the province of Quebec and the state of Louisiana. Based on what you know, how would you explain this?

   ____

   ____

3. In your own life's experiences, how has movement affected you?

   ____

   ____

   ____

4. Using events from history in the United States and Canada, identify three examples in which certain groups of people have had to move.

   ____

   ____

   ____

5. Using events from history in the United States and Canada, identify three examples of goods being moved.

   ____

   ____

   ____

6. Using events from history in the United States and Canada, identify three examples of the spread of different ideas or information.

   ____

   ____

   ____

Challenge Activity: How did the events of September 11, 2001, impact the movement of people, goods, money, and ideas?

____

____

Name: ______________________ Date: ______________________

United States and Canada: *Movement—Geoquest*

Directions: Look at items found in your home or classroom. Find 20 items from as many different countries as possible.

	ITEM	BRAND NAME	MADE IN ...
1.	Refrigerator		
2.	Automobile		
3.			
4.			
5.			
6.			
7.			
8.			
9.			
10.			
11.			
12.			
13.			
14.			
15.			
16.			
17.			
18.			
19.			
20.			

Using a copy of the world map on page 115 and the numbers of each item above, put the number of each item in the country in which it was made. Draw arrows from each country to your hometown.

United States and Canada: *Regions*

The fifth theme of geography is **regions**. The world is not only divided into political regions, bound by borders; it can also be divided into regions that are defined by physical and human characteristics. For example, the world could be broken up into language regions. One could map all the Spanish-speaking countries in the world. The world can be segmented into climate regions such as warm and tropical areas or cold and snowy areas.

A **region** can be defined as an area that shares common physical or human characteristics with another place. For example, the United States and Canada share the same language, as well as sharing the Rocky Mountains.

Geographers can map regions by crops, products, and industry as well. For example, northern California and Newfoundland are both fishing regions.

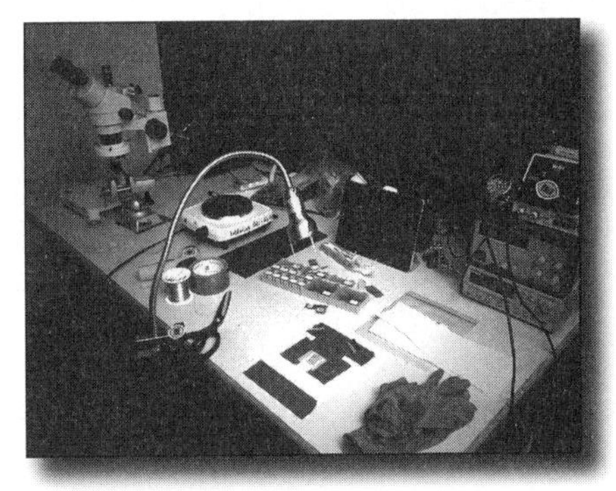

Regions can change over time. With the movement of people from one area of the world to another, regions can be redefined. War, famine, and natural disasters are causes for people to move and in the process spread language, religion, and culture. Missionaries, political leaders, and the media are influential in changing or expanding cultural regions. When looking at the United States and Canada, some physical regions are the Corn Belt, the Cotton Belt, the dairy regions, and the wheat regions. Some industrial regions include timber, oil, coal, automobiles, and electronic technology.

The U.S.-Canada Border

Name: ______________________ Date: ______________________

United States and Canada: *Regions—Activity Page*

Directions: Compare and contrast the United States and another country of your choice. What similarities do they share? What differences are there? Compare and contrast religions, language, government, climate, landforms, clothing, holidays, homes, animals, foods, transportation, and traditions.

	United States	**Country:** ________	**Similar or Different?**
Religion(s)			
Language			
Government			
Climate			
Landforms			
Clothing			
Holidays			
Homes			
Animals			
Food			
Transportation			
Traditions			

Name: ______________________________ Date: ____________________

United States and Canada: *Regions—Geoquest*

Directions: Think about what physical or human characteristics the United States and Canada have in common. Using the map below, show these regions. Make a map key to indicate the colors for each region. Some physical characteristics could be animals, plants, mountains, water, and climate. Some human characteristics could be language, religion, education, government, housing, clothes, and holidays.

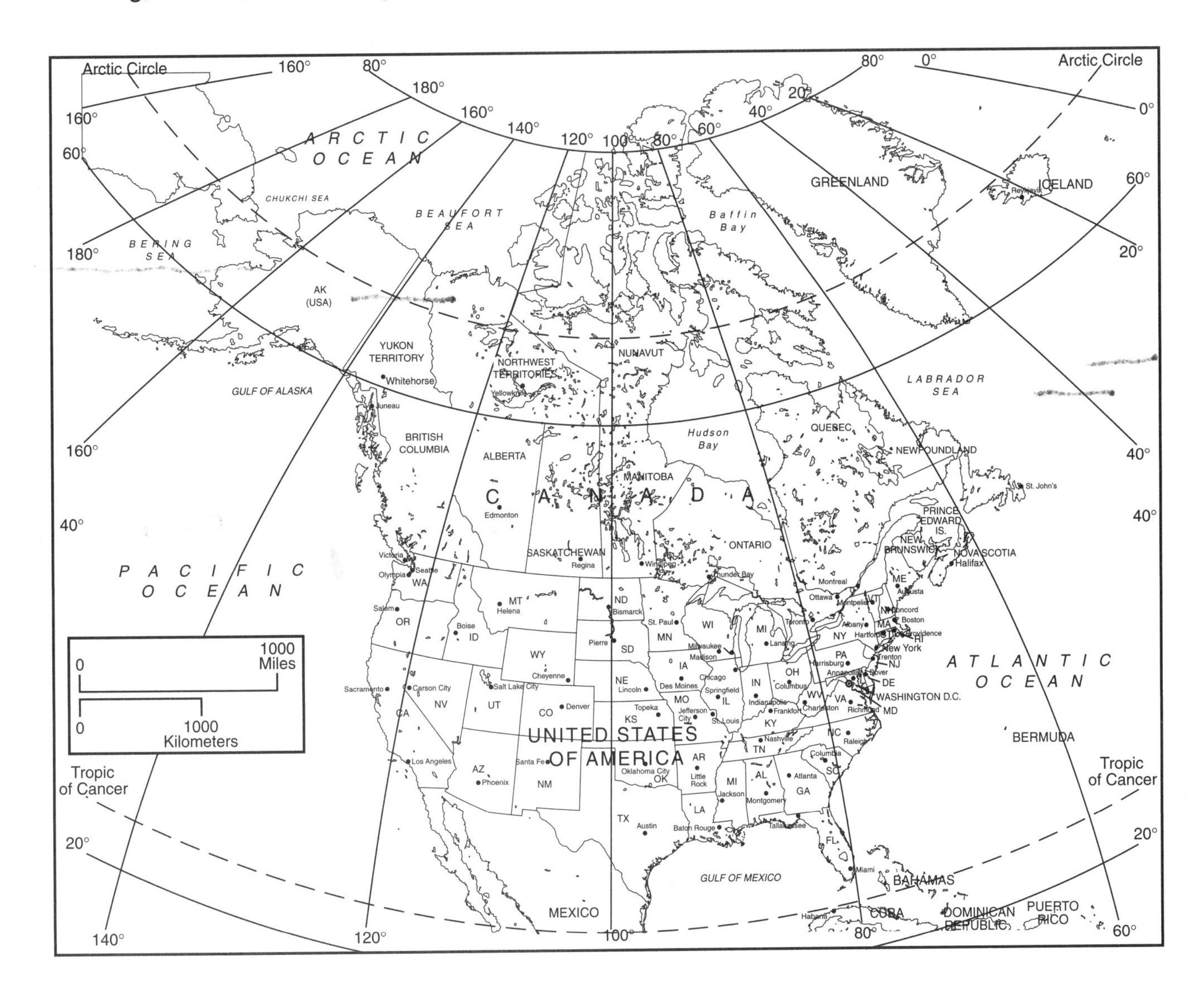

Central and South America: *Location*

The study of location is the first theme of geography. When looking at Central and South America, their locations affect the nature of the people who live there. The countries of this region are located from along the equator to approximately 30 degrees North latitude and 55 degrees South latitude. The region extends from 35 degrees West longitude to 115 degrees West longitude.

The South American continent is bordered to the northeast by the Atlantic Ocean and to the north by the Caribbean Sea. To the west, South America is bordered by the Pacific Ocean; to the southeast is the South Atlantic Ocean. Central America includes the countries from Mexico in the north to Panama in the south. The Pacific Ocean is on the west, and the Gulf of Mexico and the Caribbean Sea are on the east.

South and Central America were the homes to several ancient Native American civilizations. These civilizations had established very detailed cultures, languages, religions, and practices. But with the exploration of the Europeans, the "white man" brought diseases that the natives had not previously seen. As a result, many ancient civilizations were wiped out because the natives got sick and died. Others were conquered by the Europeans through force.

The relative location of Central and South America has helped many of its countries develop. Located to the south of the United States, Central America has been closely involved with the history of the southwestern United States through colonization attempts and civil wars. First Spain and then Mexico claimed much of the southwestern United States during the colonial period and up to the war between Mexico and the United States. The United States has also become involved in the affairs of many of the Central and South American countries. For example, the United States helped Panama become an independent nation because it contained the best location for a canal that would connect the Atlantic and Pacific Oceans.

The equator runs through several of South America's countries. This location allows many South American countries to support various types of agriculture. Anything from fruits and nuts to coffee, grains, and livestock is produced on the continent.

Central and South America: *Place*

When looking at the second theme of geography, place, we need to look at the special characteristics that make Central and South America unique on the earth.

The physical characteristics of the South American region are generally tropical. Sou' America is located on and near the equator. Jungles and wetlands dominate the landscape e country of Brazil is located near the center of the continent. The Amazon River snakes t' ough the country, watering the interior of the continent and providing nourishment to the lands. The Andes Mountains run the length of the continent along the western coast fr north to south. The countries o gentina and Chile reach to the uthernmost tip of South Ameri These countries have a variet climates such as subtropical sert, steppe, highland, and mari west coast.

Santiago, Chile

ntral America is an isthmus, a rrow body of land connecting o larger landmasses. Mexico is the largest country in this region and has a warm, arid climate. Mountains dominate the central and western parts of Central Ameria.

Major cities in South America an entral America, such as Buenos Aires, Argentina; Santiago, Chile; Rio de Janeiro, Brazil; Mexico City, Mexico, are modern, bustling areas of commerce and culture. They are am some of the largest cities in the world.

The human characteristic the Central and South American region a iverse. Many of the citizens can trace thei ckgrounds back to Native Americans a r their European conquerors. Languag he region is primarily Spanish; however language of Brazil is Portuguese. The jor religion is Catholicism. Housing and d ings vary with the location. In the Andes Mountains, huts are built for warmth. In the Amazon Jungle, dwellings are built to provide air circulation and protection from the sun. Just as the dwellings are diverse, so are the physical and human characteristics.

A floating village on the Amazon River

Name: ______________________ Date: ______________

Central and South America: *Place—Questions*

Directions: Answer the following questions.

1. Describe the physical characteristics of South America. Tropical

2. Describe the physical characteristics of Central America.

3. Describe the human characteristics of South America.

4. Describe the human characteristics of Central America.

5. Define *isthmus*. A narrow body of land connecting to 2 larger land masses.

Name: ______________________________ Date: ____________________

Central and South America: *Place—Geoquest*

Directions: Choose one or more of the following activities to complete. You may use any variety of traditional and/or digital media technology available to demonstrate your knowledge.

1. Select a specific country in the South or Central American region and prepare maps showing population, natural resources, rivers, and cities.

2. Prepare a travelogue on the Amazon River. Describe the journey and what you might see and do.

3. Prepare a multimedia presentation on the Andes Mountains, showing where they are located, who lives in the surrounding areas, plants and animals in the surrounding areas, and famous climbing expeditions and their dates.

4. Create a crossword puzzle or a word find with ten cities from Central and South America.

5. Using the Internet, find and print articles relating to Central and South America.

The Jari River is a tributary of the Amazon River.

Central and South America: *Interactions*

Assessing the interactions between humans and the environment is the third theme of geography. Humans become adapted to living in a specific environment and, in turn, try to change the environment to better fit their needs.

South America is the fourth largest continent and the fifth largest in population. The terrain of South America affects the lives of its peoples.

Nearly half of all South Americans make their living by farming. Most farms are quite small and can produce only enough food for the families that own them. Most of these people use old-fashioned ways of farming, with no machinery. There are huge modern farms and ranches; however, they are owned by a small number of wealthy people. Some of these farms are larger than some of the states of the United States. The farms grow huge quantities of coffee, cacao, wheat, sugar, bananas, rice, and other foods.

Coffee beans are a major crop in Central and South America.

The rain forests, hot, moist, and thick with vegetation, might seem to be an ideal place for farming. But when it is cleared of trees, the jungle soil loses the important nutrients crops need to grow. Much of the rain forest has been cut down in a futile attempt to create farmland. However, after only a few years of farming, the land becomes unproductive.

Life in the big cities of South America is much like life in the cities of North America. There are tall, modern buildings, airports, and busy streets. The cities of South America are some of the fastest growing in the world. Many cities face overcrowding, poverty, and crime. However, many of the Native Americans who live outside the cities still live the way their ancestors did.

Zipline tours have developed as an adventurous way to see the natural beauty of the jungle canopy.

In the mountains, the population must be prepared for severe, cold weather. In the desert regions, residents must adapt to the extremely hot and dry climate.

Many areas in Central America are becoming known for using the environment to their advantage. They encourage tourists to come visit and view their ecological resources, such as rain forests, beaches, mountains, exotic animals, and favorable climates. This is called **ecotourism**. It brings money into the economy of Central America and encourages the people to preserve the environment.

Name: ______________________________ Date: ____________________

Central and South America: *Interactions—Questions*

Directions: Answer the following questions:

1. How do most South Americans make a living? ______________________________

2. List six major agricultural products of South America.
   a. ______________________________
   b. ______________________________
   c. ______________________________
   d. ______________________________
   e. ______________________________
   f. ______________________________

3. When the rain forests and jungles are cleared of trees, what happens to the soil?

4. Only a few wealthy people control much of the land. In your opinion, why do you suppose this is?

5. How is ecotourism good for the environment?

Name: ______________________ Date: ______________

Central and South America: *Interactions—Geoquest*

Directions: On the map below, identify the major products of South America and label where they are produced.

Central and South America: *Movement*

Movement is the fourth theme of geography, and it is the study of how people, goods, and ideas travel around the globe. Movement has been extremely important in the development of Central and South America.

Reed boats and a village on Lake Titicaca between Peru and Bolivia.

Ancestors of Native Americans crossed a narrow bridge of land between what is now Alaska and Siberia thousands of years ago. Over the centuries, the Native Americans populated all of North, Central, and South America. In the Andes Mountains, a sophisticated Native American group called the Incas thrived and created a huge empire. The lands once controlled by the Incas now make up the nations of Peru, Ecuador, and Bolivia.

Just like North America, South America was explored and conquered by Europeans after about 1500. People from Spain, Portugal, and other European countries took over the land, some of which had been inhabited by Native Americans for centuries. Many wars have been fought over the years, but the borders of many of today's South American countries have existed for over one hundred years.

Using the different waterways in South America, such as the Amazon River, the people have been able to move goods and ideas into otherwise unreachable areas of the continent. The Panama Canal, which cuts through the Isthmus of Panama, is an important water route between the Atlantic and Pacific Oceans. However, railroads and highways have not been developed in Central and South America as much as they have in the United States.

Cruise ships, pleasure boats, and cargo vessels make the journey between the Atlantic Ocean and the Pacific Ocean through the lock system of the Panama Canal.

Like North Americans, most Central and South Americans speak the language of the European country that once ruled the area in which they live. For example, Brazil was once a colony of Portugal, and today most Brazilians speak Portuguese. Many other South American countries were once dominated by Spain, and Spanish is widely spoken on the continent. English, Dutch, and French are also spoken in some areas of Central and South America. There are many Native Americans in Central and South America who still speak the languages of their ancestors.

Name: ______________________ Date: ______________

Central and South America: *Movement—Questions*

Directions: Answer the following questions:

1. What route did the ancestors of Native Americans take to get to South America?

   ____

   ____

   ____

   ____

   ____

Ancient ruins at Machu Picchu in the Andes Mountains

2. Name the Native American empire that once controlled the lands now known as Peru, Ecuador, and Bolivia.

   ____

3. List two languages spoken in the countries of South America.

   a. ____

   b. ____

4. What two bodies of water does the Panama Canal connect?

   ____

   ____

5. In what ways are the 2016 Summer Olympics in Rio de Janeiro an example of movement? What makes these games significant in South American Olympic history?

   ____

   ____

   ____

   ____

   ____

Name: ________________________ Date: ________________

Central and South America: *Movement—Geoquest*

Directions: Choose one or more of the following activities to complete. You may use any variety of traditional and/or digital media technology available to demonstrate your knowledge.

1. Prepare a time line of the history of Brazil, Mexico, or Argentina.

Woven goods from craftsmen in the Andes Mountains

2. Prepare a report on trade between the United States and Mexico.

3. Research several explorers and note their contributions to the spread of European culture to South and Central America.

4. Make a list of items in your home or school that have been made in a South or Central American country.

5. Make a list of words, traditions, or items from South or Central America that have become common parts of North American civilization.

A farmer's market stand in Colombia

Central and South America: *Regions*

Within Central and South America there are many different smaller areas and regions. The fifth theme of geography is regions. This theme deals with the characteristics in different places that are similar.

South America has many important cities. The biggest of them is São Paulo, Brazil—one of the largest cities in the world. Buenos Aires, Argentina, and Rio de Janeiro, Brazil, are also in the world's top ten in population. All three cities are very modern and have a lot of industry. If you look at these three cities on the map, you will see they all have something in common: they are all near the Atlantic coast. They all grew up around or very close to natural ports, or places where ships could safely dock.

São Paulo, Brazil

South America's largest and most populated country is Brazil. More people live in Brazil than in all other South American countries combined. Brazil is also the continent's leading industrial nation. Argentina is the second-largest South American country.

Argentina contains a large ranching region. It is one of the largest producers of beef in the world. Herds of beef cattle and sheep are raised on giant ranches. Venezuela and Ecuador are among the oil-producing nations of the world. Oil rigs drill into the earth and bring up crude oil. These two nations are the largest oil exporters in South America. An **export** is a product that is made and sent out of the country to be sold. An **import** is a product made in another country and brought in to be sold.

The Patagonia region of Argentina

Different physical regions in Central and South America include mountains, grasslands, deserts, rain forests, and coastal areas.

Nearly a fourth of all the species of animals known live in South America. As in other parts of the world, people are hunting these animals or attempting to develop the lands the animals live on, so many creatures are in danger of becoming extinct.

Soccer, or *futbol* in Spanish, is one of the world's most widely played sports. It is the national sport of several South American countries, so this could be considered a soccer-playing region.

Name: ______________________ Date: ______________

Central and South America: *Regions—Activity*

Directions: Using an almanac or the Internet, fill in the following information on the chart.

Country	Form of Government	Leader	How Chosen	How Long in Power	Political Parties	Population
Brazil						
Peru						
Argentina						
Mexico						
Colombia						
Ecuador						
Venezuela						
Panama						

Name: ________________________________ Date: ______________________

Central and South America: *Regions—Geoquest*

Directions: Choose one or more of the following activities to complete. You may use any variety of traditional and/or digital media technology available to demonstrate your knowledge.

1. Prepare a travel itinerary for a trip around Central or South America.

2. Find websites with photographs of different Central or South American countries and tourist sites. Prepare a digital presentation.

3. Prepare a series of maps showing Central and South America in the year 1800, in the year 1900, and again for today, indicating population growth changes over time.

4. Select one aspect of Central or South American culture and prepare a report for the class.

5. Prepare a Central or South American food and share it and the recipe with the class.

6. Locate and bring into class some products from Central and South America that are available in the United States.

Many Central and South American fruits and vegetables have become commonplace in the United States.

Europe: *Location*

Location! Location! Location! This phrase is used often as people try to find the best spot to live or to buy property for a business. However, this phrase can be used to describe the location of Europe as well. Europe's distinction as a continent has been argued throughout history. Is Europe a continent? The answer is both yes and no. The textbook definition of a continent is "one of the great divisions of land on the earth," and most continents can clearly be seen on a map or a globe. However, Europe looks like it could be part of Asia. Europe is actually a large peninsula that stretches westward from the main body of Eurasia, as the entire landmass is called. Because of Europe's great importance in world history, however, it is considered a separate continent.

The first of the five themes of geography, location, answers the question, "Where is it?" Location can be described in two ways: absolute and relative. You can describe Europe's relative location by identifying what is around it. Europe is bordered by the Atlantic Ocean on the west, the Mediterranean Sea to the south, and the North Sea to the north. It also has an open connection with the Black Sea by way of the Dardanelles and Bosporus. However, geographers still debate Europe's eastern boundary. Many draw the line along the Ural Mountains, the Kara and Ural Rivers, and through the Caspian Depression to the Caspian Sea. Another unique part of Europe's location is that it has the Prime Meridian at zero degrees of longitude running through Greenwich, England.

Europe is located in the northern hemisphere and is in both the western and eastern hemispheres. The absolute location of Europe spans from 35 degrees North latitude to 75 degrees North latitude and from 25 degrees West longitude to 35 degrees East longitude. The reason for the change in direction from West to East longitude is due to the Prime Meridian being at zero degrees.

Europe's location has been ideal for trade, for conquest, for war, and for the spread of people, goods, and ideas. Because of its location and access to major oceans and seaways, Europeans colonized and explored other parts of the world. This location made Europe well-known in the world.

Name: ______________________ Date: ______________

Europe: *Location—Activity*

So Where Is It?

Directions: Describe the relative location of the following European countries.

1. Iceland Under green land, under artic circle in atlantic ocean.
2. Austria Between Germany and Hungary.
3. Portugal Between ~~Spain~~ Spain and the atlantic.
4. Germany Between Poland and france.
5. Italy In the Tyrrhenian sea and ~~A~~ Adriatic.
6. Finland In the gulf of Bothnia.
7. Sweden In between Norway and gulf of Bothnia.
8. Macedonia In Between greece alBania and Balgaria.

Directions: European nations are linked not just by borders, but by railroads as well. The following locations are railroad stops. Identify the absolute location of the stops.

9. Paris, France ______________
10. Bern, Switzerland ______________
11. Madrid, Spain ______________
12. Budapest, Hungary ______________
13. Warsaw, Poland ______________
14. Brussels, Belgium ______________

Name: ____________________________ Date: ____________________

Europe: *Location—Geoquest*

Scenario: You are the manager of a talented and popular band. You are in charge of planning the band's European tour. Where would the band perform? The pilot of the private jet needs the absolute locations of the cities in order to make the flight plans for the trip. Map out the band's tour and use an atlas to identify the cities' absolute locations. There is a map of Europe below to help you. In order for this map to be geographically correct, make sure you include a map title and a map key.

Europe: *Place*

"Not just places on a map," is a phrase that truly applies to Europe. There are many places in Europe that hold historical and cultural meaning. They are not just places on a map. They have physical and human characteristics that make them special or unique.

The view across the English Channel from France to England

Physical characteristics are found in nature, but could change over time due to weather and natural disasters. An example of a physical characteristic is the Seine River, which divides the city of Paris, France, and flows across western Europe into the English Channel. The English Channel, another physical characteristic, is a body of water that lies between England and France. In the past, the Channel provided a buffer between the warring nations. Today the nations are friendly, and the Channel is easily crossed by boat or train. This improves the flow of people and goods between the two countries and the rest of Europe. Another physical characteristic is the Alps mountain range, which provides skiing for natives and visitors of Switzerland, Italy, Austria, and France. Mountains are a physical characteristic that can change over time due to weathering. The Rhine River is western Europe's most important inland waterway. Canals join the Rhine with a number of other rivers, such as the Ruhr River. A greater volume of freight is carried on the Rhine River than on any other river system in the world. Because of these physical characteristics, this part of Europe is unique.

The human characteristics of place are things that humans do to change or influence their environment. This can lead to positive or negative effects. Humans build and construct objects and structures on the earth's surface. Think of the Tower of London, Stonehenge, the Leaning Tower of Pisa, the Colosseum, and the Eiffel Tower, which are all European tourist sites. They are man-made, human characteristics that make this area of the world unique. If a person just names physical features of an area in Europe, it might sound like an area of the United States. However, the human characteristics combined with the physical characteristics make the place different from that of the United States. For example, Paris, France, in landscape, weather, climate, and its river system might look like St. Louis, Missouri, of the United States. Both cities have rivers running through them, have rural areas on the far outer rims of the cities, and have comparable climates. But if one looks at the human characteristics, such as language, money or currency, and man-made structures, you will quickly realize that the place is definitely Paris, France, and not St. Louis. Can you compare the area in which you live to an area in Europe? Compare and contrast the physical and human characteristics of place.

The Seine River and Eiffel Tower in Paris, France

Name: ______________________________ Date: ____________________

Europe: *Place—Activity*

Experience Europe!

Directions: Pick a place in Europe that you would like to visit or perhaps may have already visited. Find out what makes that place unique and convince others that it is the best place in Europe to see and experience. On the map of Europe, locate your place. Then answer the following questions.

1. Name your place and describe three of its physical characteristics.

   ____

   ____

   ____

2. How do these physical characteristics impact what Europeans do with their land?

   ____

   ____

   ____

3. Explain three human characteristics of the place. ____________________

   ____

   ____

4. How do these human characteristics make this part of Europe unique? __________

   ____

   ____

5. On your own paper, explain how these physical or human characteristics have changed over time. Why would these characteristics draw people to visit this part of Europe?

Name: ______________________________ Date: ______________________

Europe: *Place—Geoquest*

Operation Railroad!

Directions: European nations are linked not just by borders and roads, but also by an extensive passenger train system. Choose cities that you might travel to by train. On your own paper, list the cities in Europe and explain why you would want to visit them. Include the human and physical characteristics that make those cities unique and great places to visit. Use the map of Europe below to label your train stops.

Europe: *Interactions*

Geographers use the third theme, human-environment interactions, to study how people affect their environment and how the environment affects them. Humans can adapt and change in order to live well in their surroundings. They may also change their surroundings to fit their needs. This can be positive or negative.

Ski slope in Austria

In Europe, the environment plays a big role in how people live. The climate, landforms, and weather affect how people dress, how they move from place to place, how they grow crops, and what they do. For example, the climate of Europe has made it a great place to visit. Tourists visit Austria, France, and Germany just to ski in the Alps. People have changed their environment by building ski lifts and ski resorts in the Alps for tourism. Travelers go to the Cote d'Azur of France in order to enjoy the beautiful beaches of the southern Mediterranean coast. The French have interacted with their environment by building cities and resorts along the Mediterranean Sea. The climate of France has made it possible for many areas to be wine-producing regions. Some of these environmental factors have made places famous and unique compared to other spots in the world.

Even though many of these human-environment interactions would be seen as positive, many interactions can cause negative results as well. Humans building and changing their physical surroundings often means tampering with or destroying natural habitat. For example, as people tear down forests in order to build ski resorts, an animal habitat could be destroyed. It could mean forced movement for that animal, or it could become an endangered species. Also, as cities develop and people move to these areas, the amount of pollution usually increases. This is a negative effect of human-environment interaction.

Hazy skies over Athens, Greece

In Athens, Greece, air pollution is threatening health and eating away at the many ancient monuments. Controls have been placed on industry and traffic movement in the capital city. Furthermore, in Greece, a species of tortoise called Hermann's Tortoise is threatened because people have collected it for the pet trade. Rapidly increasing road traffic is another danger for the tortoise. In France, the Pyrenean Desman, a long-nosed water mole, lives in the Pyrenees (mountains in southern France), and polluted streams are threatening its survival. The Italian government is attempting to deal with city pollution by restricting traffic flow in their crowded centers. Unfortunately, industry, agricultural chemicals, and traffic have harmed the environment of northern Italy. What is the cost of humans interacting with their environment? At what point does progress negatively affect our environment? You decide.

Challenge Activity:

Research the financial crisis in Greece (starting in 2008), and propose a viable solution using your knowledge of geographical human environment and interactions.

Name: ______________________ Date: ______________

Europe: *Interactions—Activity*

Human-Environment Interactions
Positive or Negative: YOU Be the Judge!

Directions: Look at the following pictures of interactions in Europe. Read the text that goes with each picture. Then on the lines provided, indicate if you think the interaction is positive, negative, or both, and explain why. Next, find a current example of an interaction going on in your community. Finally, search in the news for a current event example of an interaction in Europe.

1.

The Kops-Stausee dam generates hydroelectricity for industry in western Austria. Hydroelectric projects on Austria's lakes and rivers provide two-thirds of the nation's electrical power.

____

____

2.

Crops are grown in central Europe in traditional strip or ribbon fields. Each year the crops are planted in a different strip. This method of farming is called rotation.

____

____

3.

The Croatian War of 1991–92 impacted the region's wineries. Despite the war, Croatians have worked to continue wine production.

____

____

Name: ______________________________ Date: ______________________

Europe: *Interactions—Geoquest*

Our Endangered World

Directions: The following inhabitants of Europe are currently endangered. Using the text on page 39 and other resources, find out why. Using the Internet, magazines, and the library, find other endangered species in Europe. Explain why they make up part of our endangered world. Explain what interactions are taking place between humans and their environment.

Endangered Species (From where?)	Reason Why Endangered	Interactions Taking Place
1. Hermann's Tortoise (Greece)		
2. Pyrenean Desman (France)		
3. Pond Bat (the Netherlands)		
4.		
5.		
6.		

Europe: *Movement*

Geographers use movement to study how people move, how ideas move, and how goods are moved. This is an important part of how cultures have developed and changed over time. Think about what our world might be like if we didn't have movement of people, goods, and ideas; our world would look so much different. People from other cultures would not come in contact with each other. Societies would not get new ideas such as technology, religious ideas, languages, government practices, and even news. Societies would only have the products that they make themselves or grow themselves. Fortunately, societies have developed many ways to move.

Europe is a perfect example of movement. Throughout Europe's history, war and various leaders have moved troops in order to take control of land and to change or move borders. For example, in 1961, the Berlin Wall was built to separate East and West Germany. This limited the movement of people, goods, and ideas. In 1989, the Berlin Wall was torn down, which gave people the freedom to move. East and West Germany became united into one reunified Germany.

The Berlin Wall once kept the people of Communist East Berlin trapped in their part of the city.

Movement can be affected by the environment and the time period in which the movement occurs. In Europe, the movement of goods would be different depending on where you are located. For example, if you are a business owner in Germany, you may use the Rhine River to transport your products. However, if you live in Barcelona, Spain, you could ship goods through the Mediterranean Sea. Movement is more difficult if you have to move goods over a body of water, such as between England and France, or if you have to move goods over a mountain range, such as the Alps. Due to the small size of the European continent, much of Europe is linked by an extensive rail system.

Since Europe is located so close to Asia and Africa and the open sea lanes to the Americas, opportunities for the movement of goods, people, and ideas in this region have been the greatest in the world. Europeans have been able to spread ideas that have shaped and influenced the modern world. The European voyages of discovery, exploration, religious missionary efforts, and colonization helped to spread European ideas, religious beliefs, traditions, and languages around the globe. North America, South America, and Australia are inhabited largely by descendants who speak European languages. Think about your family. Did anyone move from Europe to the United States? What European languages are taught at your school?

Challenge Activity:

Create a map showing the movement of borders over the past 25 years with regards to the country of Yugoslavia and it's subsequent breakup. Convert that map into a multimedia presentation to share with the class using any number of apps.

Name: ______________________________ Date: ____________________

Europe: *Movement—Activity*

Move! Move! Move!

Directions: People, goods, and ideas from Europe have spread around the world. Let's look at just how these movements affect your life. Fill in the chart below with information about you and your family and where those items, ideas, or people came from. How many of these things originated in Europe? You may need to ask your parents for some family information.

	Type of Item or Type of Person	**Country of Origin**
Language		
Government		
Clothing		
Religion		
Family or Ancestor's Name		
Traditions		
Favorite Food		
Style of house		
Cars, technology, etc.		

Name: ______________________________ Date: ______________________

Europe: *Movement—Geoquest*

BUSINESS 101

Directions: You are opening your own business. You know that Europe is a great market for your product, but you must research how to transport your product to the various countries in western Europe. Airfare for freight is too costly, so you must rely on shipping, the railway system, or trucking. As the manager of this business, create a presentation or plan to present to the board of directors. They are investors in your company, and you need their votes in order to stay in business. Choose the product, the ideal countries to sell to, the routes, and the system of transportation that would be the most beneficial to your company. Include the maps below.

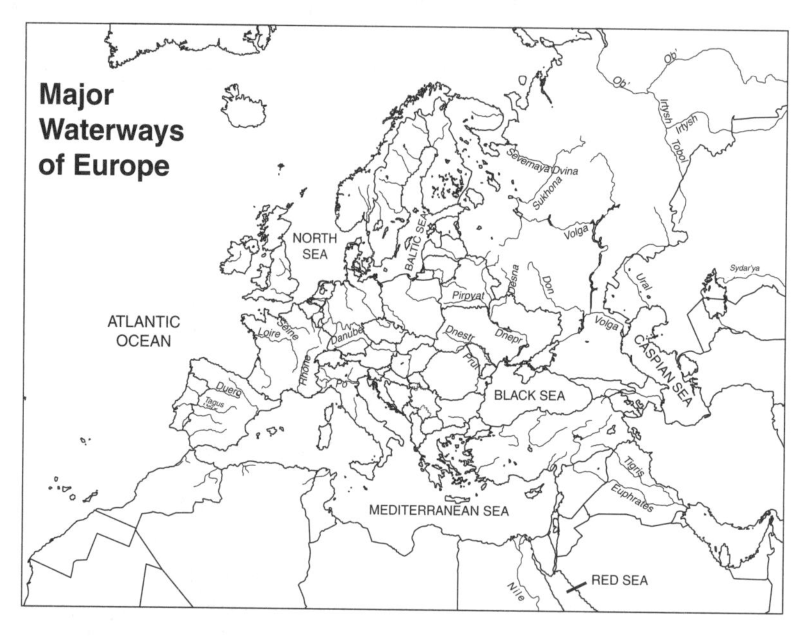

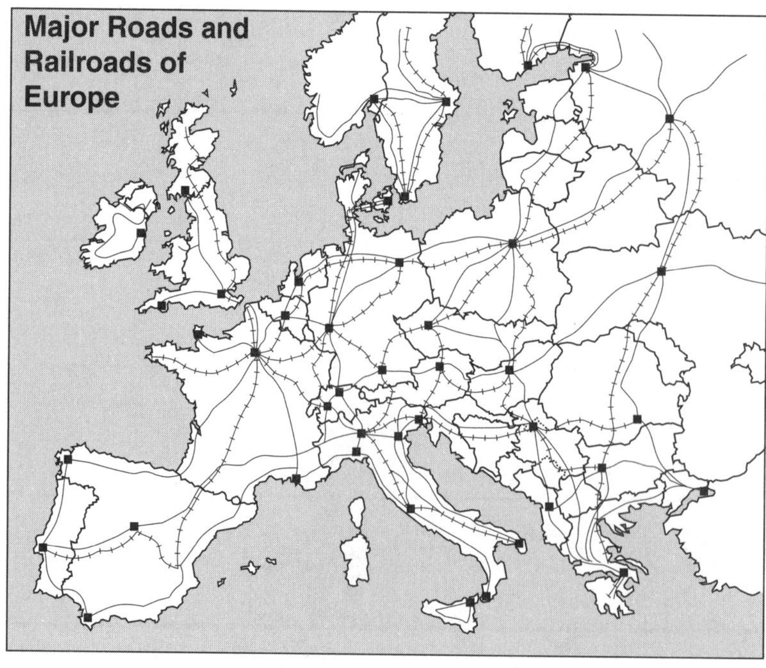

Europe: *Regions*

Geographers use the fifth theme, regions, to have a frame of reference or a way to compare and contrast areas of the world. This world can be defined according to regions. A region has certain characteristics in common that tie the area together. In Europe, there are many human and physical characteristics that areas have in common. Political boundaries are not enough. One must look beyond the borders to see what characteristics can unify Europe.

The mountains and fjords of Norway

The location of Europe has been key in developing its regions over time. It is bordered by the continent of Asia to the east, and just across the Mediterranean Sea to the south is Africa. Therefore, Europe has had much human influence in the past. Conquest, trade routes, war, and economic conditions have changed regions in Europe. Once nearly all of Europe was ruled by the Roman Empire. Napoleon and Adolf Hitler also tried to rule large empires. For many years, there was a region of communist governments in the east versus democratic governments in the west. Today Europe is made up of many independent, democratic nations, however, some regions are constantly changing their borders due to political and social unrest.

The environment also impacts Europe and divides it into regions. For example, Norway, Sweden, and Finland lie so far north compared to the southern coastal areas of Italy and Greece that they are considered a region called Scandinavia. Regions can be based on language, climate, landforms, religion, type of government, crops, industry, animals, clothing, housing, recreation, and so on. Since much of Europe is a tourist area, one can map the ski regions or the great cathedrals. Wine regions and cattle regions are examples of other regions that can be mapped. The wine or grape region is located in both western and eastern Europe. Grapes are grown in France and Germany, as well as in the dry regions of Hungary, Romania, and Bulgaria. One climate region of Europe is the Mediterranean climate region. Most of the southern part of western Europe enjoys hot, sunny summers; the winters there are usually mild and rainy.

The Mediterranean coast of France

The study of regions is essential to the unity of the world. It is important for people to look at the similarities and differences of regions and learn how to keep a balance in order for our world to function as a whole.

Name: ______________________________ Date: ______________

Europe: *Regions—Activity*

Regions Galore!

Directions: The maps below show different types of regions in Germany and Great Britain. Study both maps and answer the following questions.

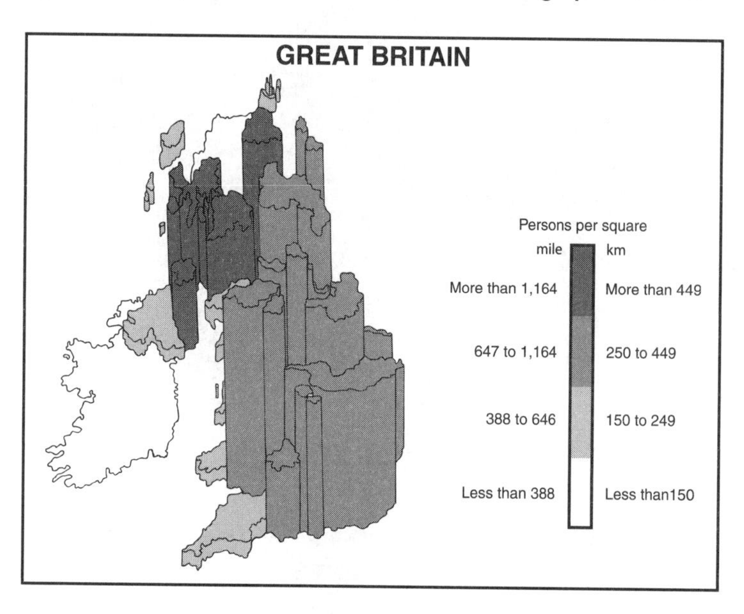

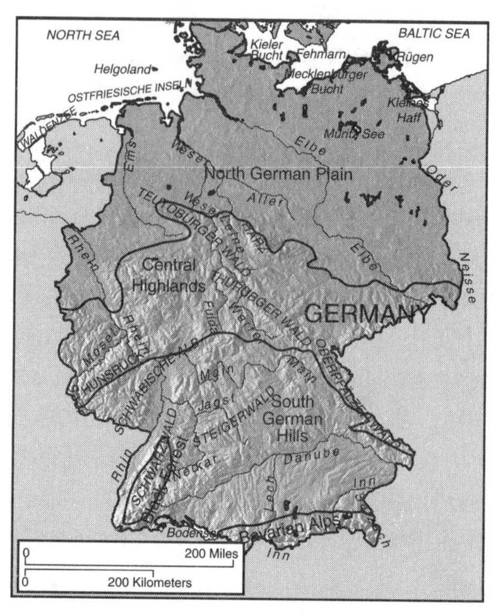

1. Describe what the map of Germany is showing. Why is this important for geographers?

   ____

   ____

   ____

2. How does this impact the people living in Germany? ______________________________

   ____

3. In what region would you expect most of Germany's agriculture to take place?

   ____

4. Describe what the map of Great Britain is showing. Why is this important for geographers?

   ____

   ____

   ____

5. How does this impact the people living in Great Britain? ______________________________

   ____

6. The majority of Great Britain is in which population range? ______________________________

   ____

Name: ______________________________ Date: ______________________

Europe: *Regions—Geoquest*

Bon Voyage!

Directions: Create a travel brochure that highlights the regions of Europe to visit. Think about the ski regions, the chocolate regions, the wine regions, the cathedrals, and so on. Then use the suitcases below to make a list of things you need as you travel through different types of climate regions represented below.

Russia: *Location*

The country of Russia and the Independent Republics that surround it make up nearly one-sixth of the earth's land. These countries span two continents—Europe and Asia. Due to the vastness and great diversity of the land and its peoples, many different climates and types of vegetation are present. Russia's absolute location is approximately 30 to 175 degrees East latitude and 50 to 75 degrees North longitude.

The traditional date for the founding of Russia by Vikings along the Dnieper River is A.D. 862. Russia was once many different kingdoms ruled by princes. The kingdoms eventually came together and were ruled by one leader. The people worked to support their leader. In 1917, Russia was a **monarchy**, a government ruled by a king or queen. In the case of Russia, the leader was called a **tzar**. The monarchy was overthrown in a revolution, and the revolutionaries established a communist form of government. **Communism** is a type of government where the state owns or controls all of the major industries, the press, and the distribution of goods and services. This country was called the Union of Soviet Socialist Republics (U.S.S.R.). In 1991, the people again led a revolution, and the Soviet Union collapsed. Today the region is known as Russia and the Russian Federation.

Russia is located in the eastern half of Europe. In Asia, it extends south to the Caspian Sea and is bordered by the Pacific Ocean on the east and by the Arctic Ocean to the North. Russian climates are typically cool because Russia is located farther north of the equator than other countries. Russia has access to the seas from the north, but in winter weather, much is frozen over. Many areas of the land, including Siberia, are barren and cannot be farmed. As a result, the population is largely located in western Russia, the part of the country on the European continent. Cities such as Moscow and St. Petersburg have large populations, while towns in eastern Russia are virtually ghost towns.

Relatively speaking, a part of the landmass of Russia is close to most areas of the world. It stretches approximately 140 degrees of longitude in the eastern hemisphere and over 30 degrees of latitude in the northern hemisphere. At one time, Russia owned the area of the United States known today as the state of Alaska.

Name: ______________________ Date: ______________

Russia: *Location—Questions*

Directions: Answer the following questions.

1. What type of government did Russia have before 1917? ______________________

2. Across which two continents does Russia span? ______________________

3. Identify the absolute location of the following cities:
   a. Moscow ______________________
   b. St. Petersburg ______________________
   c. Vladivostok ______________________
   d. Volgograd ______________________
   e. Petropavlovsk ______________________
   f. Murmansk ______________________
   g. Rostov ______________________
   h. Novosibirsk ______________________

The Winter Palace in St. Petersburg

4. Describe the relative location of the country of Russia. ______________________

5. Define *communism.* ______________________

6. Define *monarchy.* ______________________

Name: ______________________________ Date: ____________________

Russia: *Location—Geoquest*

Directions: Choose from the following activities. You may use any variety of traditional and/or digital media technology available to demonstrate your knowledge.

1. Create a travel brochure describing Russia. Include pictures and maps.

2. Prepare a report on the history of Russia from 1900–2015. Focus on the changing governments.

3. Research Russia's different ethnic groups. Prepare a chart outlining the similarities and differences.

4. Searching the Internet, find websites about Russian cities and share them with your classmates.

A knife-wielding Kabardian performer. The Kabardian people live in the northern Caucasus region between the Black and Caspian Seas.

5. Locate the Crimean Peninsula on a map. This area has been part of Ukraine. Recently, the people of the area voted to become part of Russia again. Prepare a report about why the location of this peninsula is important to Ukraine and Russia and why the people feel strong ties to Russia.

Russia: *Place*

There are many characteristics of Russia that make it a very diverse place. Place is the second theme of geography, and it looks at the special aspects that make each region unique to the earth.

Lake Baikal frozen in the winter

The physical characteristics of Russia vary from east to west and north to south. The southwestern region of Russia, on the west side of the Ural Mountains, has rolling farmlands dotted with small villages. The southern area is more temperate, with lands adjacent to the Black Sea and the Caspian Sea. Some of the areas on the sea coasts have a more Mediterranean climate with resorts and vacation areas. To the north, the landscape is barren tundra, bordering the Arctic Ocean. Although Russia is one of the largest regions in landmass, it has very little land suitable for farming and grazing. East of the Ural Mountains is the Siberian Plateau, which stretches to the Pacific Ocean. The countries of Mongolia and China border Russia to the south of this area.

During the 1900s, the Russian people developed from a serf or slave population to a more independent people. During the feudal period, princes and lords controlled large amounts of land, and the people were forced to tend the land to make a living. It wasn't until the late 1800s that this practice came to an end. With the Russian Revolution in 1917, the people were again dominated by a new form of control called communism. Since 1991, the Russian people have begun experiencing the growing pains of forming a democratic society.

The language of the region is largely Russian, and many Russians practice Russian Orthodox Christianity. Russian cities, like Moscow, are large and overcrowded. The quality of life is poor in comparison to American cities. Other cities such as St. Petersburg (formerly Leningrad until 1992) and Vladivostok are heavily populated and are major centers for commerce and culture. Life is changing dramatically in this region of the world.

St. Basil's Cathedral in Moscow

Name: ______________________________ Date: ________________

Russia: *Place—Questions*

Directions: Answer the following questions.

1. Describe the physical characteristics of western Russia. ____________________

2. Describe the physical characteristics of eastern Russia. ____________________

3. Describe the human characteristics of Russia. ____________________

4. Why were the Russian peoples largely a "slave" population until 1917?

5. Define *democracy.* ____________________

The port city of Murmansk in northern Russia

Name: ______________________________ Date: ______________________

Russia: *Place—Geoquest*

Map Activity: On the map below, label five major cities and five major rivers of Russia. Label the surrounding countries and the surrounding bodies of water.

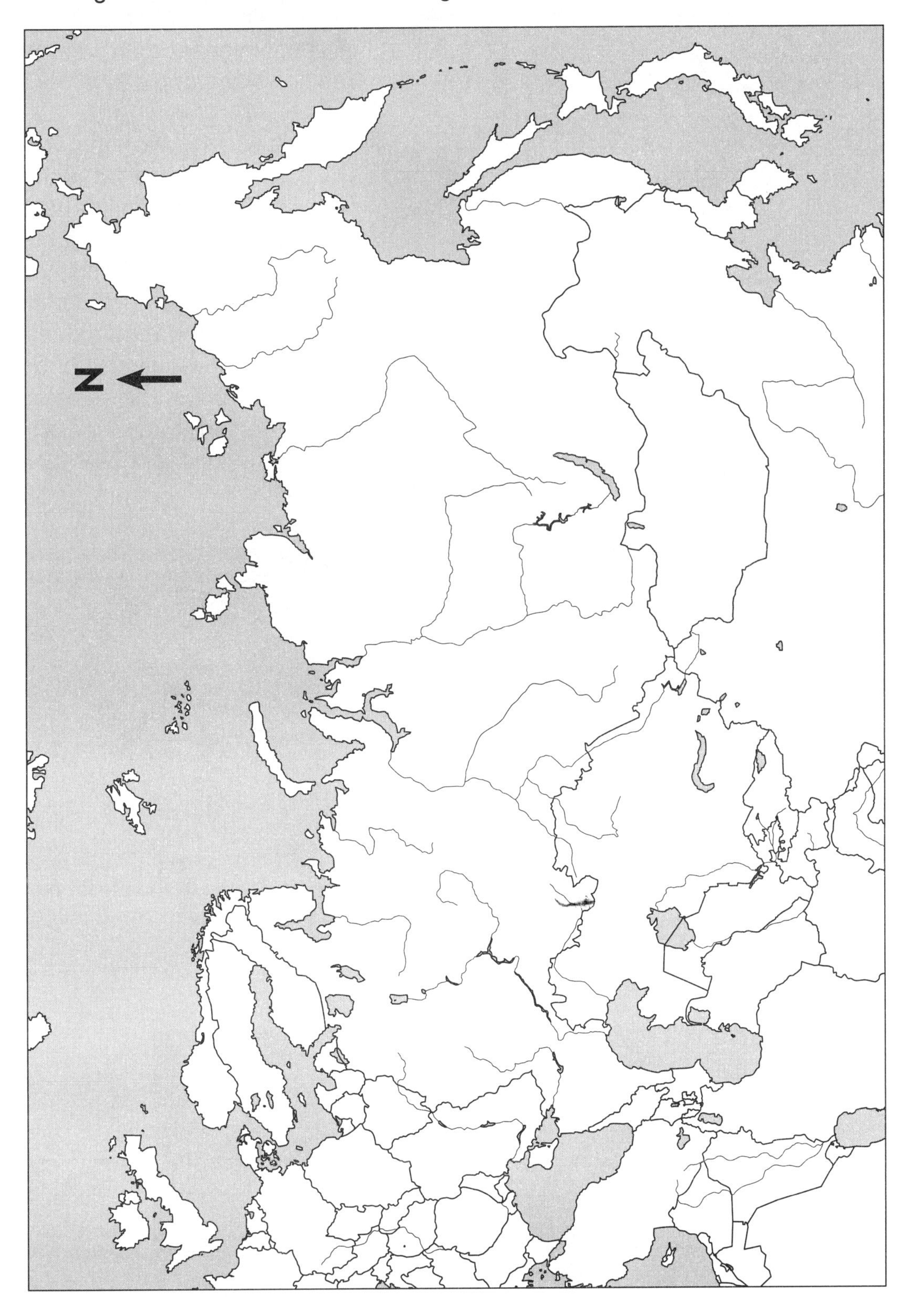

Russia: *Interactions*

Interactions between the people and the places in which they live are a key aspect in the study of geography. Human-environment interactions is the third theme of geography. How do Russians affect their environment, and how does the environment affect the Russian people?

New containment structures have been built around the old reactors of the Chernobyl nuclear power plant in Ukraine.

Russians have negatively affected their environment in many ways. The amounts of pollutants that are released into the air are damaging the ecosystem. Factories and automobiles do not have the rigorous emissions controls that are required in America. Acid rain is destroying natural forests and damaging buildings. Another example of a negative affect on the environment is the nuclear disaster at Chernobyl in the 1980s. The nuclear radiation given off when the power plant experienced a malfunction affected humans, animals, and farmlands. Russians have positively affected their surroundings since the 1990s by reducing automobile emissions and working to reduce pollution from factories.

The physical environment also affects the Russian people. Because of the severe, bitterly cold winters, the people have learned to adapt. Clothing and hats are made of warm furs. Homes are heated much of the year. Because much of the population in small villages is still agrarian, or farm-related, the people have learned to work the land and produce the crops they need to survive. Fishing in the northern regions is another industry upon which the Russian people depend for food. Many Russians vacation in the warmer southern regions near the Black Sea and the Caspian Sea. The area around Sochi, which features both mountains and seaside resorts, was shown off to the world during the 2014 Winter Olympic games.

The Nenets people live in the far north in Arctic Russia. They depend on hunting and reindeer herding to survive.

The Olymic Park in Sochi, Russia

Name: ______________________ Date: ______________

Russia: *Interactions—Questions*

Directions: Answer the following questions.

1. In what ways have Russians positively affected their environment? ______________________________

2. In what ways have Russians negatively impacted their environment? ______________________________

3. In what ways does the Russian environment positively affect the Russian people?

4. In what ways does the Russian environment negatively affect the Russian people?

Name: ______________________ Date: ______________________

Russia: *Interactions—Geoquest*

Directions: Think about your environment. How is your environment different from the following Russian city? Use an almanac or the Internet to help you. Fill in the answers on the chart below.

	Your City ______________	**Moscow, Russia**
Summer Climate		
Winter Climate		
Physical Environment		
Types of Housing		
Foods		
Clothing		
Transportation		
Crops grown		
Industries/ Products		

Russia: *Movement*

The fourth theme of geography is movement. The movement of goods, services, and ideas, both within the Russian region and with the rest of the world, have been very important to the growth of the country.

A busy port on the Dnieper River at Kherson

Throughout the history of Russia, language, religion, and culture were spread by travelers from village to village. Through the use of rivers, highways, and railroads, citizens were able to communicate with one another. There are very few major roads or railroads connecting eastern Russia with western Russia. This makes the movement of people and goods difficult. However, some of the world's longest rivers are found in Russia, providing natural transportation routes. These routes mostly run in a north-south direction.

With the beginning of communism and the Soviet Union in the early twentieth century, the government strictly limited what information the people received. The media was controlled by the government. Products were rationed, which means they were given out in limited quantities. The Russian population did not have free access to the goods and services that much of the western world was developing. As a result, the Soviet Union fell behind the rest of the world in the development of industry. Instead, the government focused much of its resources on building a strong military.

The movement of ideas has greatly affected life and government in the Russian region. During the advance of the technological era of the 1980s, the Russian people began to find out more information from the western world. Radio, television, and computers brought to the Russian people images of the rest of the world enjoying fun and relaxing ways of life. As a result, Russians became more active in demanding those same opportunities. In 1991, the Soviet Union was dissolved, and Russia was reinstated as a country.

A Russian diesel locomotive

Name: ______________________________ Date: ____________________

Russia: *Movement—Questions*

Directions: Answer the following questions.

1. How did movement affect the establishment of government in Russia in 1917?

____

____

____

2. In what ways did movement affect the downfall of government in Russia in 1991?

____

____

____

3. What impact did modern technology have on the Soviet Union in the 1980s?

____

____

____

4. In your opinion, why do you think the Soviet government limited the amount of information, goods, and services that the people could use?

____

____

____

5. Instead of spending money and resources on services for the people, the Soviet government used the money in what industry?

____

Name: ______________________ Date: ______________________

Russia: *Movement—Geoquest*

Directions: Using the Internet, research the following questions.

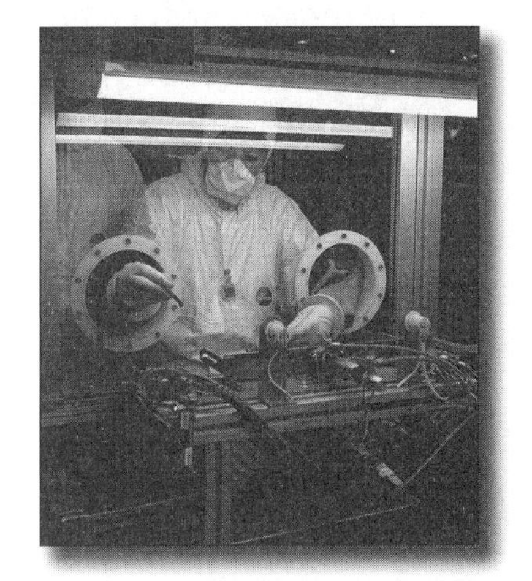

1. Prioritize the top three man-made products of Russia.
   a. ______________________
   b. ______________________
   c. ______________________

2. Prioritize the top three natural products of Russia.
   a. ______________________
   b. ______________________
   c. ______________________

3. Compare the current population of Russia to that of 25 years ago. What changes do you notice and why? ______________________

4. What is the current Gross National Product (GNP) of Russia?

   ____

5. What is the GNP of the United States? How does it compare to that of Russia?

   ____

   ____

   ____

   ____

Russia: *Regions*

A region is defined as an area that shares common physical and human characteristics. The study of regions is the fifth theme of geopraphy.

The Russian region is defined by culture, language, religion, and government, among other things. These are some of the human characteristics that make up Russia. In Russia, the main language that is spoken is Russian; however, many younger Russians also speak English. Recently, the people in the Crimean Peninsula voted to leave Ukraine and become part of Russia. This was largely because most of the peope spoke Russian and felt more a part of the Russian culture. However, the people of eastern Ukraine wanted to adopt a lifestyle closer to the other European countries.

The Russian Orthodox Church is known for its elaborate artwork, including icons showing religious figures.

Religion is predominately Russian Orthodox Christianity, which was the only church allowed to operate by the Communist Soviet government. Since the fall of communism, many other religious denominations have grown in Russia as well. The Russian people have a rich history of expressive art, and religious art is found throughout Russia.

Different physical regions exist within the Russian nation. The Ural Mountains are in the west central region. The Black Sea, Caspian Sea, and the Aral Sea are in the southwest. The southwestern region is the most conducive to farming. The mountains and the northern regions are extremely cold and barren. The Altai Mountains border Mongolia to the south. On the east is the Pacific Ocean, Sea of Okhotsk, and the Kamchatka Peninsula. The extreme northern region is the coastline of the Arctic Ocean. A majority of the population is located in the western region with the capital city of Moscow near the center of this region. Just east of the Ural Mountains is the Siberian Plateau, a large barren mass of land.

Russia is defined by its many differing physical and human characteristics.

The Khibiny Mountains of Russia

Name: ______________________ Date: ______________

Russia: *Regions—Activity*

Directions: Compare and contrast the United States and Russia. What similarities do they share? What differences are there? Compare and contrast religions, language, government, climate, landforms, clothing, holidays, homes, animals, foods, transportation, and traditions.

	United States	Russia	Similar or Different?
Religion(s)			
Language			
Government			
Climate			
Landforms			
Clothing			
Holidays			
Homes			
Animals			
Foods			
Transportation			
Traditions			

Name: _______________________________ Date: _______________________

Russia: *Regions—Geoquest*

Directions: Think about the physical or human characteristics of the Russian region. Using the map below, show these characteristics. Make a map key to indicate the colors for each characteristic. Some physical characteristics could be animals, plants, mountains, water, and climate. Some human characteristics could be language, religion, education, government, housing, clothes, and holidays.

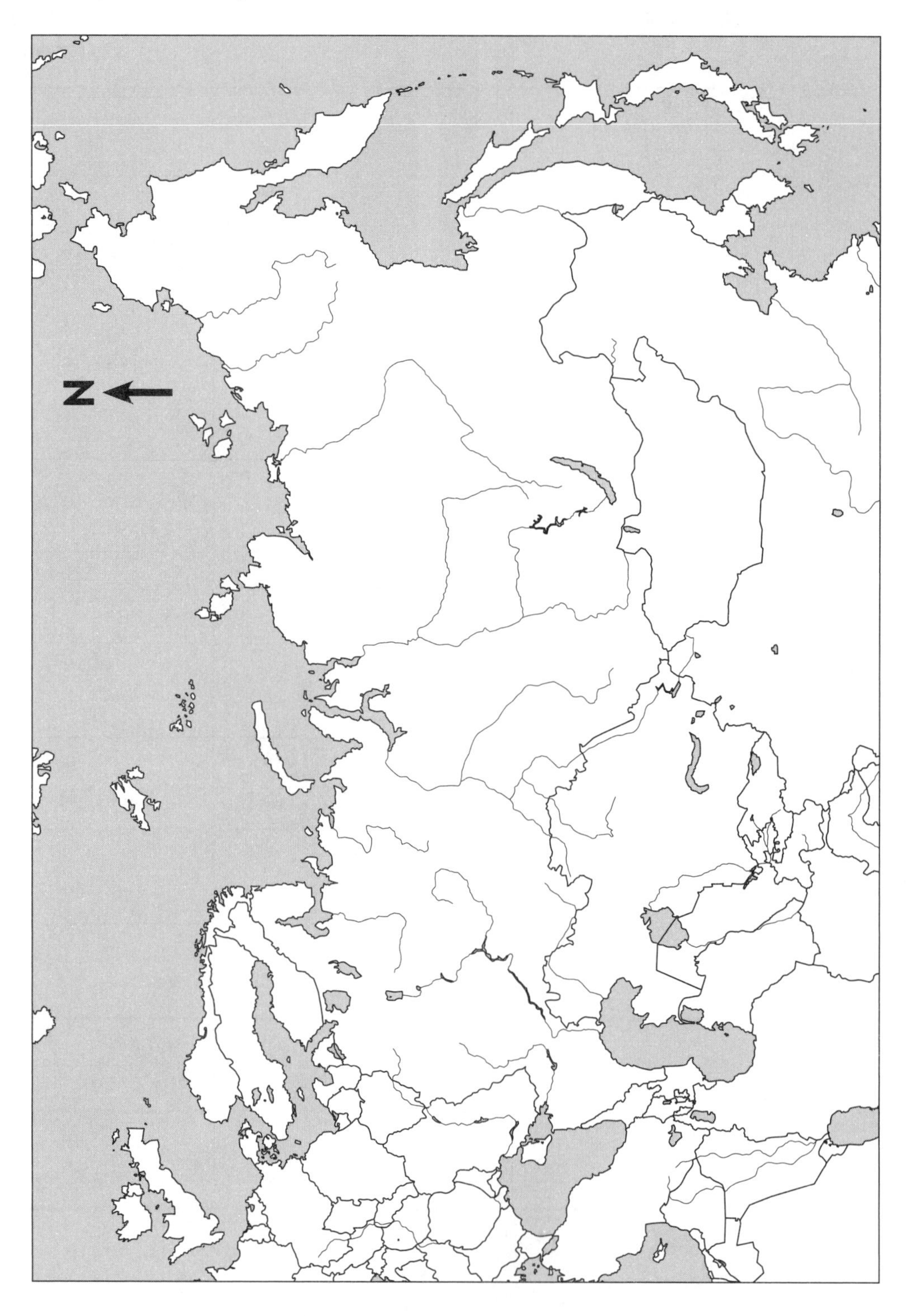

Southwest Asia and North Africa: *Location*

The first geographic theme, location, is especially important to this region of the world, North Africa and southwest Asia. This area is where three major religious homelands are located. This area has incredible access to major trade routes and waterways. It has also been a hot spot, or region of conflict. Therefore, access to certain seas has also meant strategic military advantage to certain nations. Think about the location: the Mediterranean Sea is in the center of the area, with Europe and parts of Asia to the north, the Atlantic to the far west, and access to the Red Sea, the Persian Gulf, and the Indian Ocean on the east.

Location means the relative as well as the absolute location of a place. **Relative location** is where something is in relation to other places. This is almost like giving directions using landmarks. The description of waterways that surround southwest Asia and North Africa is part of the region's relative location. **Absolute location** is the geographic coordinates of a place on the earth's surface. This is done by giving the exact latitude and longitude of the location of a place on the globe. For example, one could look at a map and determine that this area of the world lies from 10 degrees North latitude to about 40 degrees North latitude and from 20 degrees West longitude to about 80 degrees East longitude. The Prime Meridian is at zero degrees longitude and cuts through western Africa.

The Sahara Desert also serves as a landmark to identify the location of these nations. It is used as a border at times and as a frame of reference, especially because it is the largest desert in the world. It covers about one-third of Africa. The Sahara Desert extends across the northern part of the continent from the Atlantic Ocean to the Red Sea. The Sahara is bounded on the north by the Atlas Mountains of Algeria, Tunisia, and Morocco. The desert goes eastward into Egypt and includes parts of Mauritania, Mali, Algeria, Morocco, Tunisia, Libya, Niger, Chad, Senegal, Sudan, and Western Sahara. This area of the world also has many other deserts, such as the Syrian Desert as well as the huge desert in Saudi Arabia called the Rub' al Khali, which is referred to as the Empty Quarter. How does location impact the lives of the people? How does a desert location impact life?

The Sahara Desert in Libya

Name: ______________________________ Date: ____________________

Southwest Asia and North Africa: *Location—Activity*

Directions, Please

Part I Directions: The map below shows North Africa and southwest Asia. There are five countries listed below. Describe their relative locations.

1. Jordan ____

____

2. Tunisia ____

____

3. United Arab Emirates ____

____

4. Yemen ____

____

5. Libya ____

____

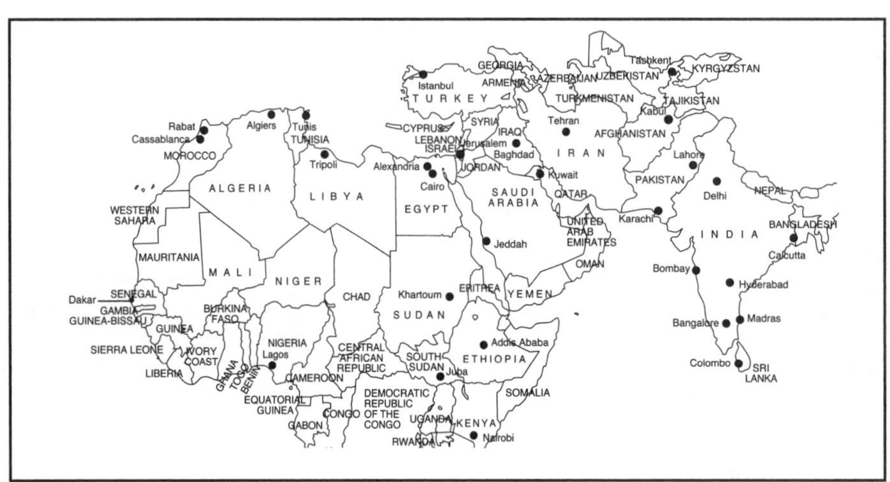

Part II Directions: Using an atlas, give the absolute location for the following places in North Africa and southwest Asia.

6. Algiers, Algeria ______________________
7. Istanbul, Turkey ______________________
8. Cairo, Egypt ______________________
9. Khartoum, Sudan ______________________
10. Baghdad, Iraq ______________________
11. Casablanca, Morocco ______________________
12. Jerusalem, Israel ______________________

Name: ____________________________________ Date: ______________________

Southwest Asia and North Africa: *Location—Geoquest*

War on Terrorism

Directions: In 2001, in order to prepare American troops for the War in Afghanistan (and later in Iraq), the troops had to be trained for desert conditions. Since that region of the world is considered a political hot spot and has so many deserts, much planning is needed for this training.

Imagine that you are working for the U.S. Department of Defense and are in charge of the planning. Complete the map below. This map should include possible sites for barracks, supply lines, and troop and equipment transports. For this activity, you will need to locate and label the deserts in the area. You may use small dots to symbolize sand, or you may shade in the desert areas. Create a map key and a title for your map.

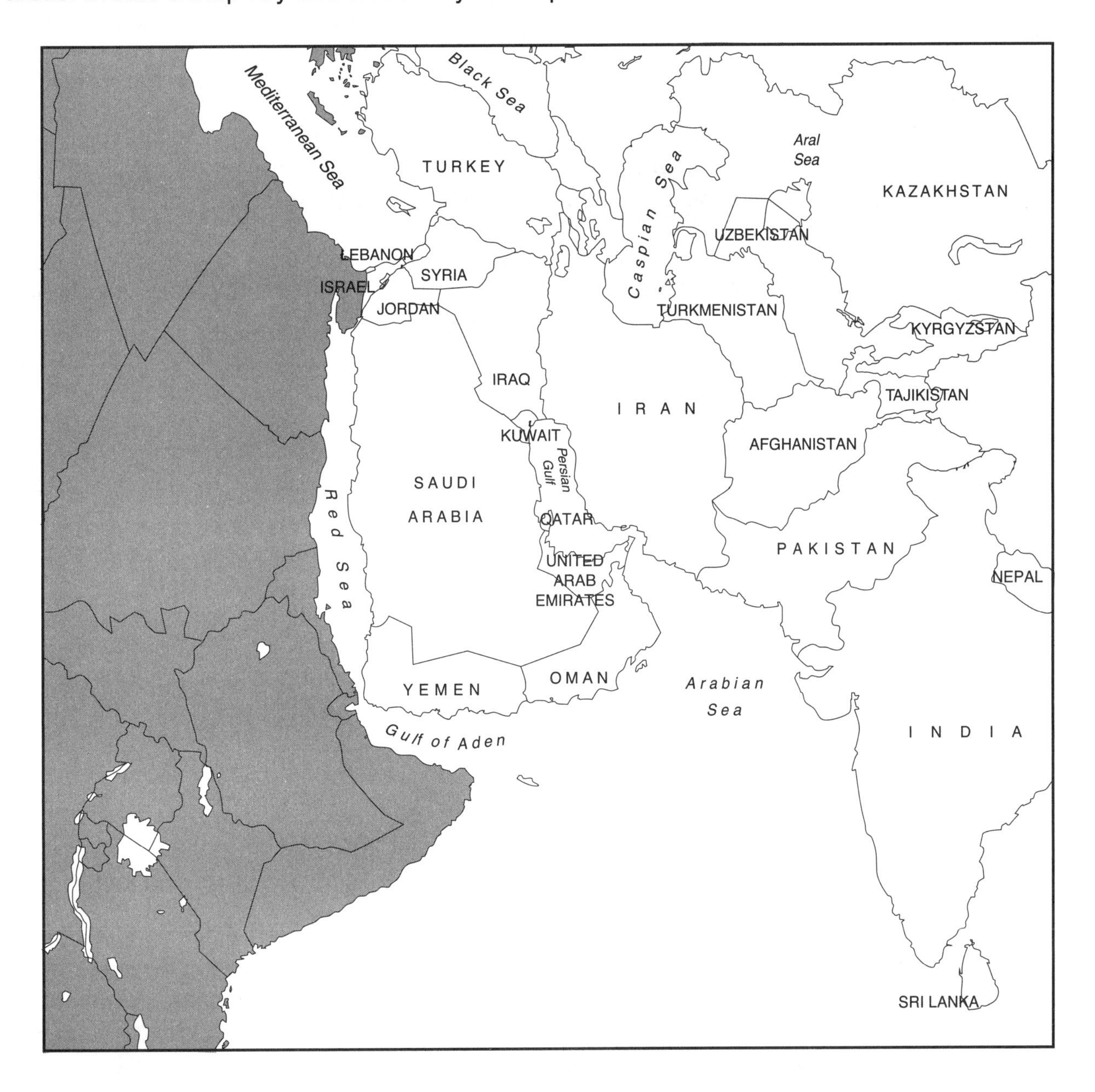

Southwest Asia and North Africa: *Place*

Step Pyramid built for King Djoser

Place, the second theme of geography, helps define a particular location. It explains what makes the place unique or different from other places on the globe. Place is made up of human characteristics and physical characteristics. **Human characteristics** are man-made structures, such as buildings, dams, bridges, roads, and schools. Human characteristics also include language, religion, and government. In northern Africa and southwest Asia, the pyramids of Egypt or the mosques would be human characteristics that make this area of the world unique. For instance, the Step Pyramid was built for King Djoser of Egypt in 2650 B.C.; it stands just south of Cairo, Egypt. The oil tankers and oil-drilling equipment of Kuwait, Jordan, Iraq, Iran, and many other countries are other human characteristics. The Suez Canal was built by people to provide an important open seaway for trade, conquest, and transportation. However, the body of water itself would be a physical characteristic. The natural environment, including natural disasters, climate, weather, plants, and animals, makes up the **physical characteristics** of place.

In order to have an accurate picture of a particular place, a person must study physical and human characteristics together. For example, if a geographer looked only at physical characteristics and discovered that a particular place has a hot and dry climate, a desert, and is near a body of water, a person could not tell if the geographer is describing the state of Arizona or the nation of Saudi Arabia. Both locations match the physical characteristics. Arizona has desert areas and is near the Colorado River, while Saudi Arabia is largely desert and is bordered by the Red Sea on the west and the Persian Gulf on the east. However, if one looks at human characteristics, the place becomes more unique. Details such as the Arab language, the government being ruled by a king, oil production, and the Muslim religion describe the human characteristics of this place. Is the place Arizona or Saudi Arabia? The answer is clearly Saudi Arabia.

Masmak Castle in Riyadh, Saudi Arabia

It is important to consider that physical and human characteristics can and will change over time. For example, weathering and the forces of wind, sand, and water have had an effect on the pyramids of Egypt. Also, the Sahara Desert, a physical characteristic of northern Africa, is changing. The desert is getting bigger due to desertification. Other changes that impact physical and human characteristics are hurricanes, tornadoes, droughts, or the force of water on rock. Human characteristics can change as well. Think of what war has done to places all over the world. In the Persian Gulf War of 1991–92, oil fields and oil refineries were set on fire. Bombs and land mines can change physical and human characteristics. Humans can tear down forests or build new businesses, causing changes throughout the landscape. What changes have occurred in your area that have impacted the physical and human characteristics?

Name: ____________________ Date: ____________________

Southwest Asia and North Africa: *Place—Activity*

Where in the World Are You?

Directions: Choose a country in the area of northern Africa and southwest Asia. Then find out what makes the country unique. Identify the physical and human characteristics of the nation. Choose a partner or small group to share your clues about your country and ask them whether you are describing a human or a physical characteristic. Then see if they can guess which country is being described. Refer to the map below to help you get started.

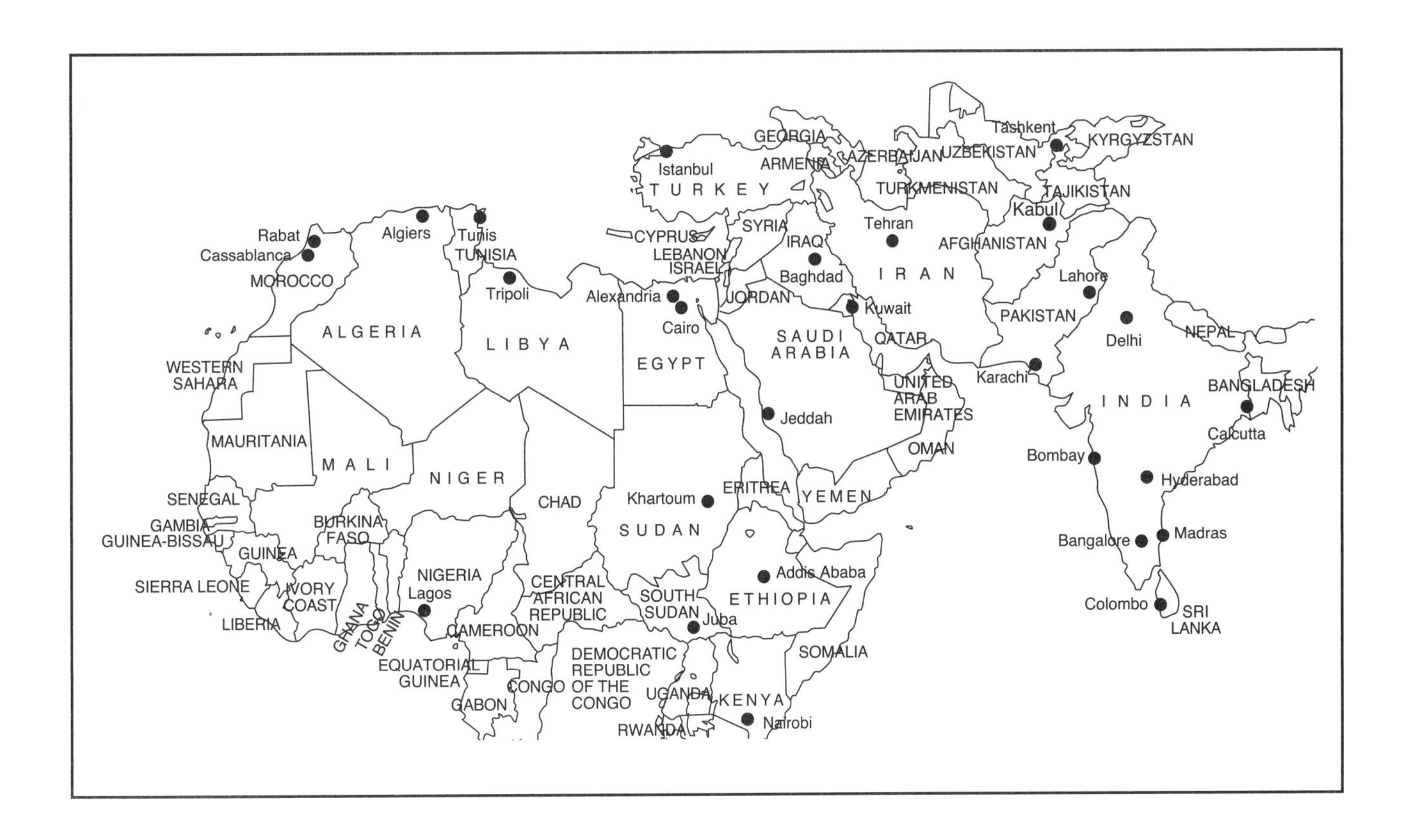

Name: ______________________________ Date: ____________________

Southwest Asia and North Africa: *Place—Geoquest*

Country Collage

Directions: Search the Internet to collect articles, pictures, or drawings to represent physical and human characteristics of place. You could focus on one country or several countries from northern Africa and southwest Asia and create a collage (either on paper or electronically). On the left side of the collage, put all physical characteristics. On the right side of the collage, put all human characteristics. If students put their collages together, it will make a great geo-mural of place.

Southwest Asia and North Africa: *Interactions*

Human-environment interactions show others how people live. Geographers use this third theme of geography to study what interactions take place in order for people to survive in their surroundings. This area of the world involves unique interactions between humans and their environment. Their climate, location, and desert regions make it essential for the people to interact with their environment in certain ways. Many of the people living in northern Africa, in the Sahara Desert, or in the Middle East have to live in certain homes and dress a certain way. Many people dress in white cotton layers because of the hot and dry conditions. Their homes often have very small windows in order to avoid letting in too much of the sun's hot rays.

A necessary human-environment interaction in the Middle East and northern Africa has been how to get much-needed water into dry, desert areas. Libya, a country located in northern Africa, has three main deserts within its borders: the Sahara Desert, the Libyan Desert, and the Surt Desert. The people of Libya have had to develop ways of getting water. The picture at the right shows massive trenches being dug for the Great Manmade River project. The plan's aim is to improve the supply of fresh water to coastal towns. It involves carrying water by pipeline from beneath the desert to the towns. In Kuwait, spectacular water towers dominate the skyline of Kuwait City. The water towers are part of desalination plants that turn seawater into fresh water by removing the salt. Fresh water is scarce in Kuwait, and these plants are its main source.

The Great Manmade River project

Kuwait Towers in Kuwait City

Mining and oil drilling are interactions that occur in northern Africa and the Middle East. These interactions are positive because the people can use oil, phosphate, and other mined materials for their own countries and as exports. These interactions are also negative because of the pollution and political conflicts that happen as a result. In the country of Bahrain, there is less oil than most Persian Gulf nations, yet their oil refinery is one of the biggest and most modern in the world. However, as they burn the gases from the oil wells, the resulting pollution blackens the skies of Bahrain. Once again, we must consider how these human interactions affect our precious environment: clean air, clean water, and the preservation of habitats.

Name: ______________________________ Date: ______________________

Southwest Asia and North Africa: *Interactions—Activity*

Directions: Since oil drilling is an excellent example of human-environment interaction in the world, use an article in the news, from the Internet, or other resources that focuses on the importance of oil and Kuwait. Then answer the questions below.

1. Describe whether you think of oil drilling as a positive or negative human-environment interaction. Why?

   ____

   ____

   ____

   ____

2. How important is this interaction to our nation's economy and to the world?

   ____

   ____

   ____

3. How much oil comes from Kuwait? What other countries provide oil?

   ____

   ____

   ____

4. Describe the process of drilling oil. What long-term effects does this process have on the environment?

   ____

   ____

   ____

   ____

   ____

Name: ______________________ Date: ______________

Southwest Asia and North Africa: *Interactions—Geoquest*

Who Depends on This Interaction Anyway?

Directions: Many people do not realize the importance of humans interacting with their environment. So we will prove them wrong. Research oil exporters and importers. On the map of the world, make a map key that shows a color representing which countries **import** (buy or take in) oil and then a different color representing which countries **export** (sell and send out) oil. In this way, everyone will be able to see how necessary this interaction is to the world economy. Make sure to give your map a title. Ask your parents to help you record how much your family depends on this interaction each week.

Oil refinery in Egypt

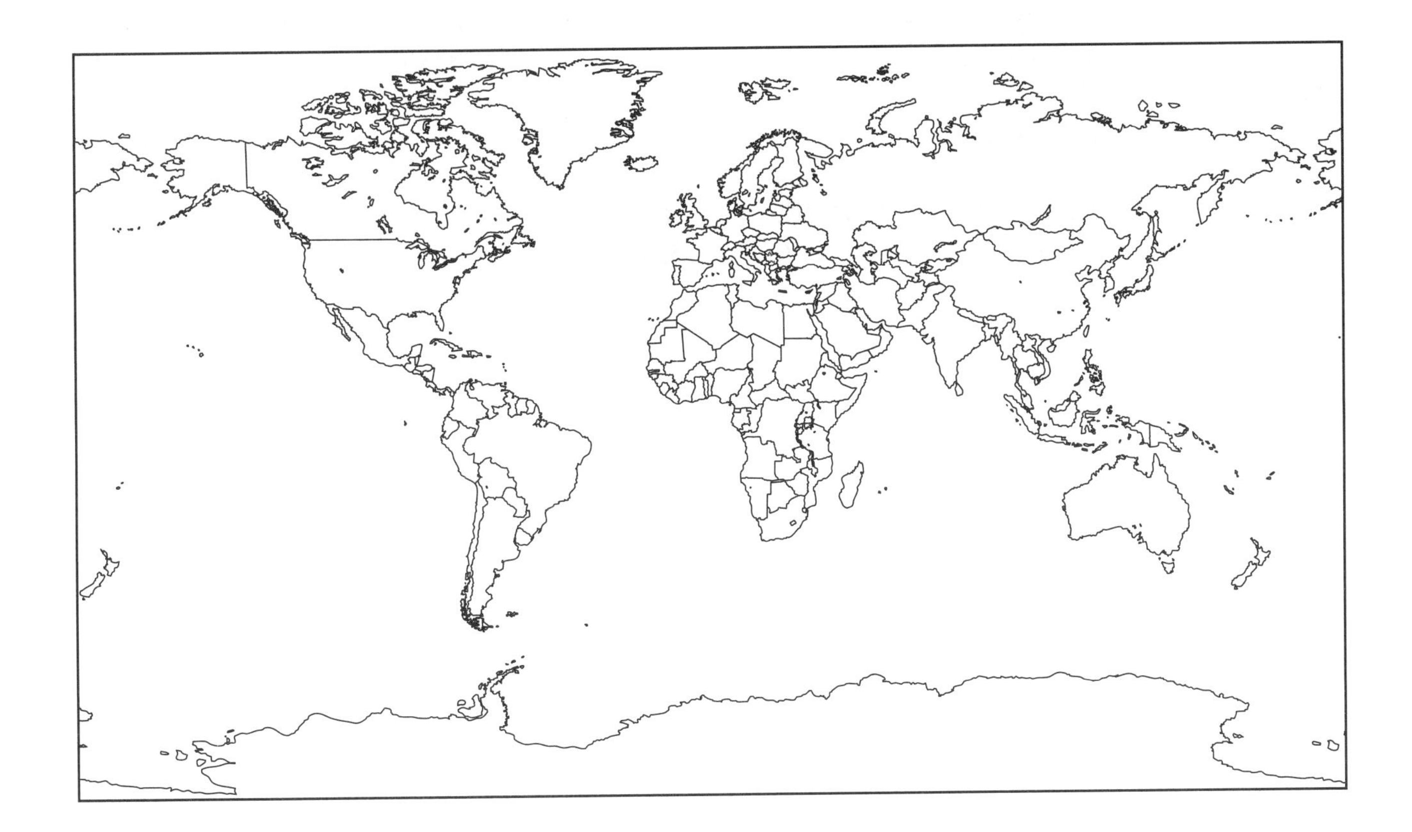

Southwest Asia and North Africa: *Movement*

Everything is on the move! Think about it. Historically, most people decided to settle near water. Why? So they would have a source or supply of water to drink, cook with, bathe in, graze animals by, irrigate crops with, or to travel on. The river systems, lakes, and oceans were their passageways to move people, goods, and ideas. This theme of movement is the fourth theme of geography. This is the study of how things move and why things move.

The moving of people, ideas, and goods can change how people live. Many nations rely on other nations for certain natural resources, raw materials, or finished products. People need to move products or goods in order to trade and earn money and get the things they need and want. People also move. They may move and actually relocate, or they may move just to visit family or to make a holy journey for religious reasons. Ideas move as well. Imagine what life was like without our modern-day technologies. It used to be more difficult to move people, goods, and ideas. People moved ideas by writing things down or by word of mouth. Missionaries were used to move religious ideas and spread them across countries. War and conflict would result when a type of government spread to a different land. Think about borders moving in this area of the world. For instance, consider the movement of Israelis and Palestinian Arabs in southwest Asia or the region considered the Middle East. Control of this area has moved from one group of people to another. People have fled as refugees from this area to move away from the conflict, while others have moved into the region as settlers.

North Africa and this area of southwest Asia have the ideal location for movement. The Mediterranean Sea lies to the north, the Atlantic Ocean to the west, the Suez Canal and Red Sea are in the middle, and the Persian Gulf is to the east. Most of the best-developed transportation and communication systems are found in the eastern Mediterranean area. Transportation and communication in the rest of North Africa and the Middle East are only partially developed. Roads and highways connect major cities with towns, oil fields, and seaports. Iran, Turkey, Morocco, and Egypt are the countries with the most extensive road systems. There are few railroads in North Africa and the Middle East. Rail lines connect major interior cities with one another and with port cities. Turkey has the most railroad mileage in the region; however, many people still travel on foot, by camel, bicycles, donkeys, and horses.

Modern technology helped with the movement of ideas during the Arab Spring uprisings that began in December 2010 in northern Africa and the Middle East. Protestors organized and got their message out using cell phones, the Internet, Facebook™, and Twitter™. Protests throughout the spring and summer of 2011 and continuing today have led to changes in the governments of Tunisia, Egypt, Lybia, Yemen, Lebanon, Sudan, Bahrain, and many other countries. Fighting is still going on in Syria, and there is unrest in other countries in the region.

If movement did not exist, there would not be any influence from other cultures, and people would be very isolated from the rest of the world. Fortunately, people in all parts of the world seem to enjoy exchanging goods and ideas. They also frequently move themselves for business and personal reasons.

Name: ______________________________ Date: ____________________

Southwest Asia and North Africa: *Movement—Activity*

Directions: The Suez Canal has been extremely important to the movement of people and goods throughout the history of this area. Find an article about the Suez Canal in an encyclopedia or on the Internet and answer the following questions.

1. When was the Suez Canal opened? ______________________________
2. Describe the location of the Suez Canal and why it is so important to this area of the world.

   ____

   ____

   ____

   ____

   ____
3. Give three examples of movement from the article you found.

   ____

   ____

   ____
4. How has the Suez Canal improved trade routes? ______________________________

   ____

   ____

   ____
5. How has conflict impacted the Suez Canal? ______________________________

   ____

   ____

   ____

Challenge Activities:

Activity One: Imagine there were no cell phones or social media in 2010 when the Arab Spring uprisings began. What form of communication would you have used to get your message out? Would the uprisings have had the same impact? Why or why not?

Activity Two: Choose one of these countries: Egypt, Lebanon, Libya, or Syria. Write an informative/explanatory essay to compare and contrast the government of the country before and after the Arab Spring uprisings.

Name: ______________________________ Date: ____________________

Southwest Asia and North Africa: *Movement—Geoquest*

A STRUGGLE FOR PEACE

Directions: This area of the world has had much military movement, control issues, conflict, and movement of borders. For decades, there has been an ongoing struggle for peace. The maps below show how the borders have moved. The United States has been involved in trying to negotiate peace in this area. Read the information with the maps and then choose one of the following options.

1. The Israelis and Palestinian Arabs are calling on the United Nations for help. Write a UN resolution and come up with a strategy to promote peace in this area.
2. Watch the news, gather newspaper articles, and present a newscast on the latest events in northern Africa and southwest Asia.
3. You be the reporter. Create a video newscast that tells both sides of this story.
4. Interview someone who is from this area of the world. Ask him or her what life is like in the Middle East and what he or she thinks of the issues.

1920 British Mandate: During World War I (1914–1918), the Turkish Ottoman Empire controlled Palestine and most of the Middle East. In exchange for Arab support in the war, Britain offered to back Arab demands for independence once the war was over. However, Britain also supported the creation of a national homeland for the Jews in Palestine. In 1920, the League of Nations gave Britain a **mandate** (authority to rule) over Palestine. During the 1930s, many European Jews moved to Palestine to escape persecution by Germany's Nazi government.

1947 UN Partition Plan: Demand for a Jewish state grew after World War II (1939–1945). Both Arabs and Jews called on Britain to keep its promises to them. Britain appealed to the United Nations (UN) for help. In 1947, the UN proposed a partition plan that split Palestine into two independent states, one Jewish and one Arab. The Jews in Palestine agreed to the plan, but the Arabs opposed it. On May 14, 1948, Israel proclaimed itself an independent Jewish state. The next day, armies from neighboring Arab countries attacked it.

1967 After the Six-Day War: Israel gained territory after defeating Arab armies in the 1948–1949 war. War broke out again in 1956, 1967, and 1973. In the 1967 Six-Day War, Israel took control of the Sinai Peninsula and the Gaza Strip from Egypt and the Golan Heights from Syria. Israel also took the West Bank and East Jerusalem from Jordan. Israel returned the Sinai Peninsula to Egypt in 1982. In 1993, Israel and the Palestinians signed the Oslo Accord, which called for Israel to return the Gaza Strip and Arab towns in the West Bank to the Palestinians in exchange for peace.

Southwest Asia and North Africa: *Regions*

Places that share common characteristics are tied together as a region. This is the fifth theme of geography. A **region** is defined as a group of places bound together by one or more similar characteristics. The characteristics can be physical or cultural (human).

Geographers study regions in order to gain insight into places and even to help people in the future. Regions can change over time due to conflict, climate, weather systems, or changes in government, economic conditions, or technology. For example, the desert region of northern Africa and southwest Asia is changing. The Sahara Desert is becoming larger, and this could cause more problems for the people who live there. Mali, for example, is dealing with this issue of desertification. Unfortunately, governments usually realize too late the effects of clearing forests for building and agriculture.

North Africa and southwest Asia have been called the Middle East by geographers. This area's location on the globe, its religions, its language, its physical environment, and even some governmental issues make this area of the world a unique region. One could map this area as an Arabic language region, an oil region, or a desert region. If a geographer maps the religious regions of the Middle East, it is considered home to three major religions: Judaism, Islam, and Christianity. The location of this area is ideal for access to the Mediterranean Sea, the Red Sea, and the Persian Gulf. This region of the Middle East is also known as a world hot spot due to the history of religious homelands, instability, and ongoing conflict. The Middle East has been a place of war, terrorism, and many unsuccessful attempts at lasting peace. Furthermore, the political regions of this area have changed over time due to conflict. The movement of borders has changed the political regions. Think about Israel and the land of the Palestinians. How many times have the political regions changed?

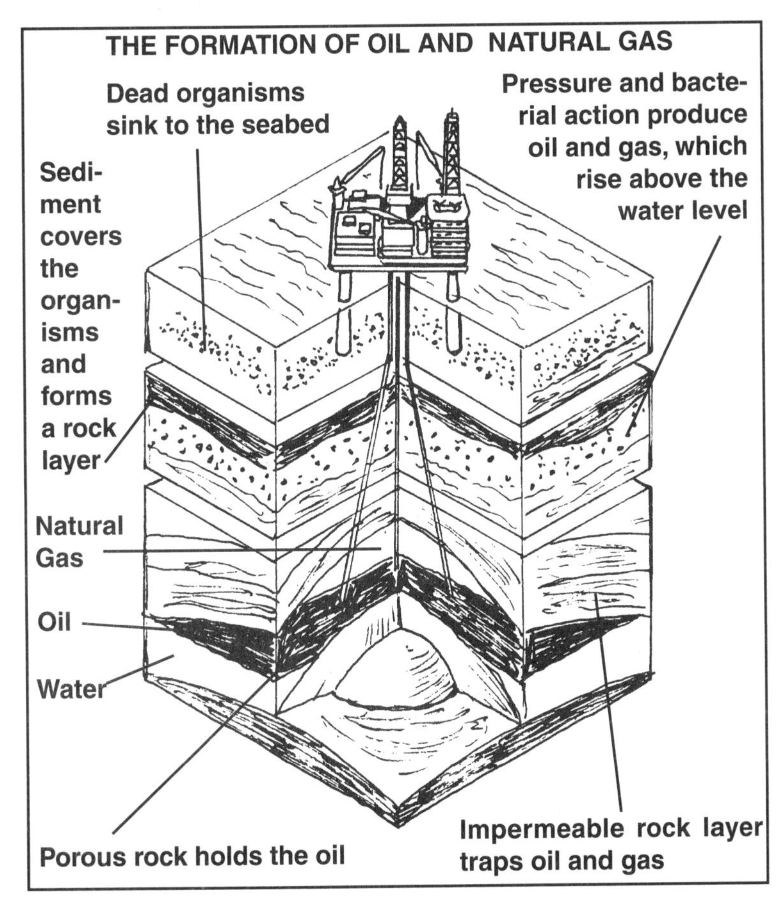

Geographers, political analysts, historians, and current world leaders will continue to study this region of the world. Since this region is important for the exporting of oil, many countries depend on the Middle Eastern oil nations. It is a region that shares common characteristics, but many problems as well. This region is in the world news and headlines often and can definitely impact the future of our world.

Name: ______________________________ Date: ____________________

Southwest Asia and North Africa: *Regions—Activity*

Hot Topic: Oil

Directions: Since oil is a resource that Americans need, the topic of oil should be hot! It is important to many parts of the world. Read the article on Kuwait and answer the following questions on your own paper.

Kuwait, an oil-rich nation in the Middle East, draws most of its income from a single source—oil. Kuwait has used most of its oil wealth to support its welfare system and modernize the country. The government has also given money to other Arab countries and non-Arab countries in Africa and Asia.

Kuwait, as a member of the Organization of Petroleum Exporting Countries (OPEC), has sometimes used its valuable oil to influence world affairs. In 1967, during the Six-Day War, Kuwait cut off oil shipments to the United States and other Western countries that supported Israel. The Arab-Israeli War, in 1973, led to other Arab nations and Kuwait stopping oil shipments to the United States and the Netherlands. In 1987, Iran attacked Kuwaiti oil tankers in the Persian Gulf because of Kuwait's decision to support Iraq in its eight-year conflict with Iran. In 1990, Saddam Hussein, the Iraqi president, ordered an invasion of Kuwait in order to acquire its oil wealth. The Persian Gulf War, in 1991, was fought to liberate Kuwait from Iraq. Iraqi troops damaged almost half of Kuwait's 1,300 oil wells. The troops set the oil wells on fire, which produced a thick black smoke and severe air pollution. The United States and a coalition of other nations helped Kuwait and brought about a successful end to the war. Afterwards, Kuwait began the huge task of repairing its oil storage, refining, and transportation facilities. However, Kuwait expected to reach its 1.5 million-barrel-a-day OPEC production quota before the end of 1992.

1. How did oil form in the Middle East region? (See diagram on page 75.)
2. Why has oil been a source of conflict in the Middle East?
3. How has Kuwait used oil as a weapon against the United States?
4. Why is the Persian Gulf important?
5. What is OPEC?
6. Why did the United States help Kuwait during the Persian Gulf War?
7. Find current event examples that connect to this region of the world.

Name: ______________________________ Date: ______________________

Southwest Asia and North Africa: *Regions—Geoquest*

Map It!

Directions: The Middle East region is the birthplace of three major religions: Judaism, Christianity, and Islam. Research where in the world Judaism, Christianity, and Islam are practiced. Then choose a color for each religion, include a map key and a map title, and map these religions below on the world map.

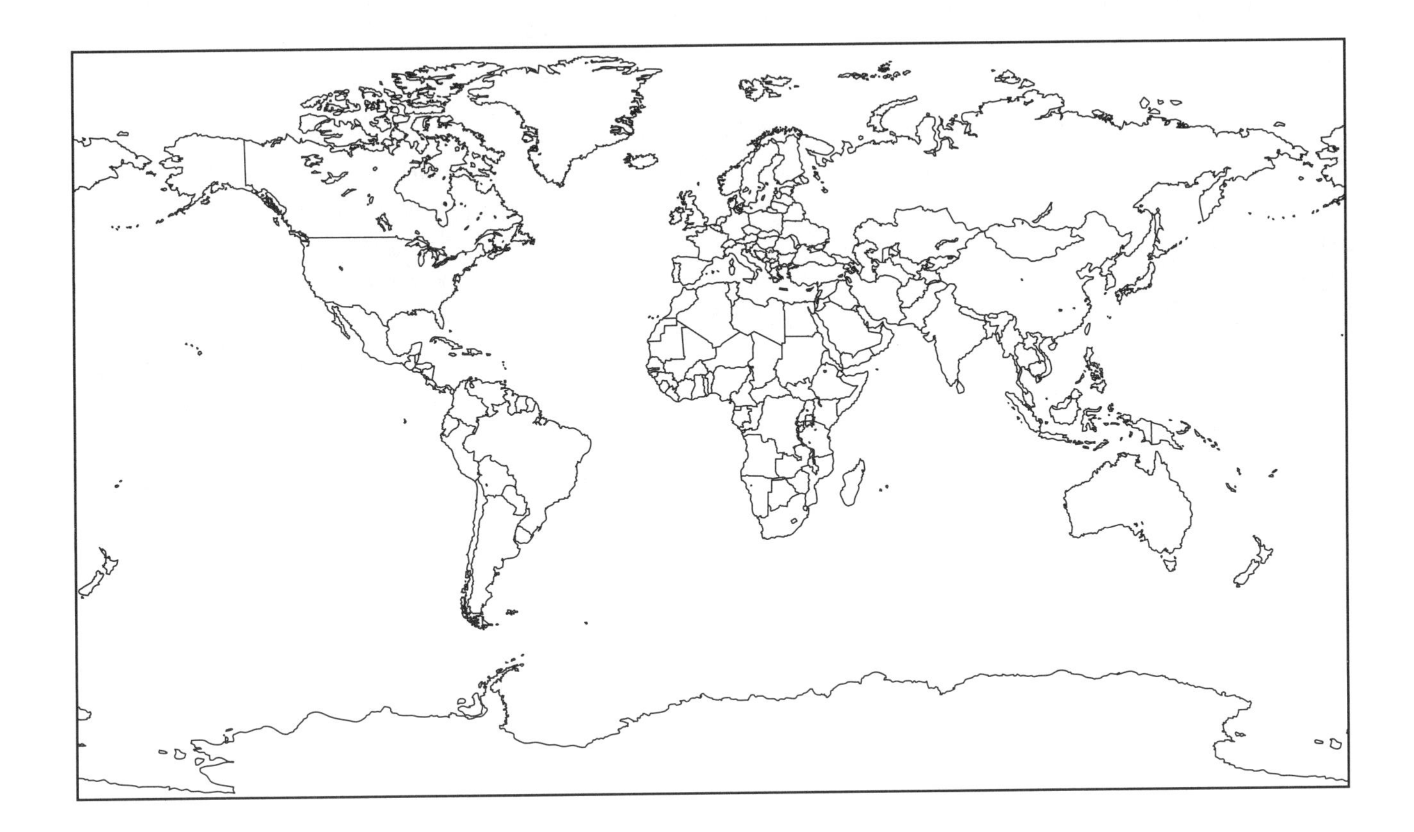

Africa, South of the Sahara: *Location*

What difference does location make anyway?

The absolute and relative locations of the Cape of Good Hope

The location of Africa or where you are within the continent of Africa has been historically important for many reasons. In our world's early history, Africa's location was vital for trade and exploration. In fact, Africa is considered to be the location of the birthplace of man. In 1967, a jawbone fragment of one of man's early ancestors was discovered in Kenya. Anthropologists believe this bone to be 5,000,000 years old.

Where is Kenya located? First, you need to locate Africa on the world map. Africa's location can be described by its relative location and its absolute location. Africa's **relative location** can be described as being bound by the Mediterranean Sea, the Red Sea, the Gulf of Aden, the Indian Ocean, and the Atlantic Ocean. Africa is surrounded by water with the Equator and the Prime Meridian cutting through it. Kenya is on the eastern side of Africa on the Indian Ocean, and it is bordered by Somalia, Ethiopia, South Sudan, Uganda, Tanzania, and Lake Victoria. The **absolute location** of Africa would be its exact address on the earth's surface: the latitude and longitude coordinates. The absolute location of Africa is approximately 38 degrees North to 35 degrees South latitude and from 18 degrees West to 52 degrees East longitude. The absolute location of Kenya is centered at approximately 0 degrees latitude and 36 degrees East longitude.

Geographers have used the Sahara Desert to break up the continent or divide it into two parts. The Sahara Desert region of the northern part of Africa is much different from the forested mining regions of the southern part of Africa. The region from the Central African Republic and all those countries south make up the region of Africa, south of the Sahara Desert. Since the Sahara Desert is expanding or going through **desertification**, over time the desert region could get larger. The differences between the northern part of Africa and the area south of the Sahara Desert have caused geographers to study them as two separate entities. Has this situation ever happened in the history of the United States? Is one part of the United States separated because of different characteristics or physical features?

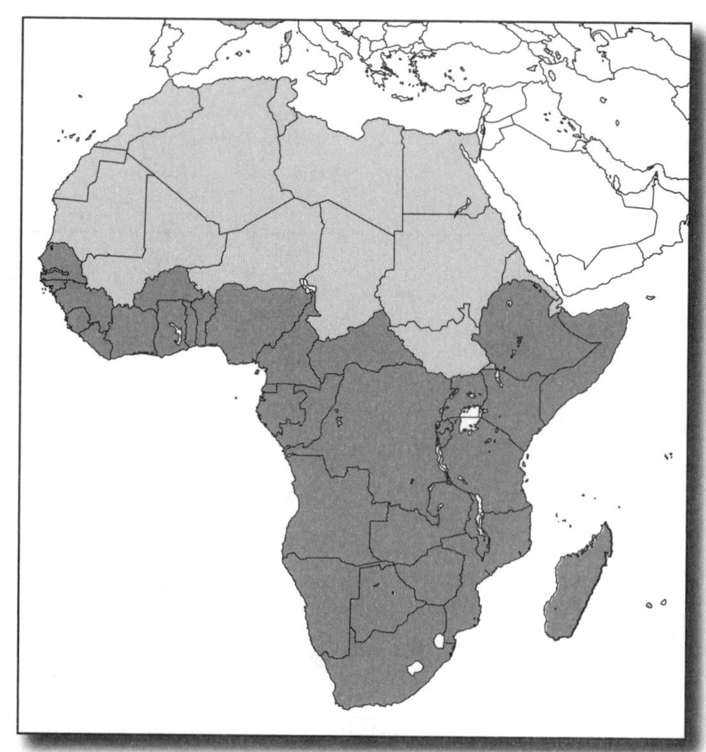

Name: ______________________ Date: ______________

Africa, South of the Sahara: *Location—Activity*

Directions: Using the maps below, the Internet, and an atlas, answer the following questions about location and the nations of Africa.

1. Describe the relative location of Lake Victoria. ______________________
2. Describe the relative location of the Drakensberg Mountains. ______________________
3. Describe the relative location of Madagascar. ______________________
4. Describe the relative location of Lake Malawi (Lake Nyasa). ______________________
5. Describe the relative location of Somalia. ______________________
6. Identify the capital located at 19 degrees South and 47 degrees East. ______________________
7. Which capital lies closest to the 20 degrees East longitude line? ______________________
8. Which capital lies closest to the 10 degrees South latitude line? ______________________
9. Give the absolute location of the Cape of Good Hope. ______________________
10. Give the absolute location of Lusaka, Zambia. ______________________

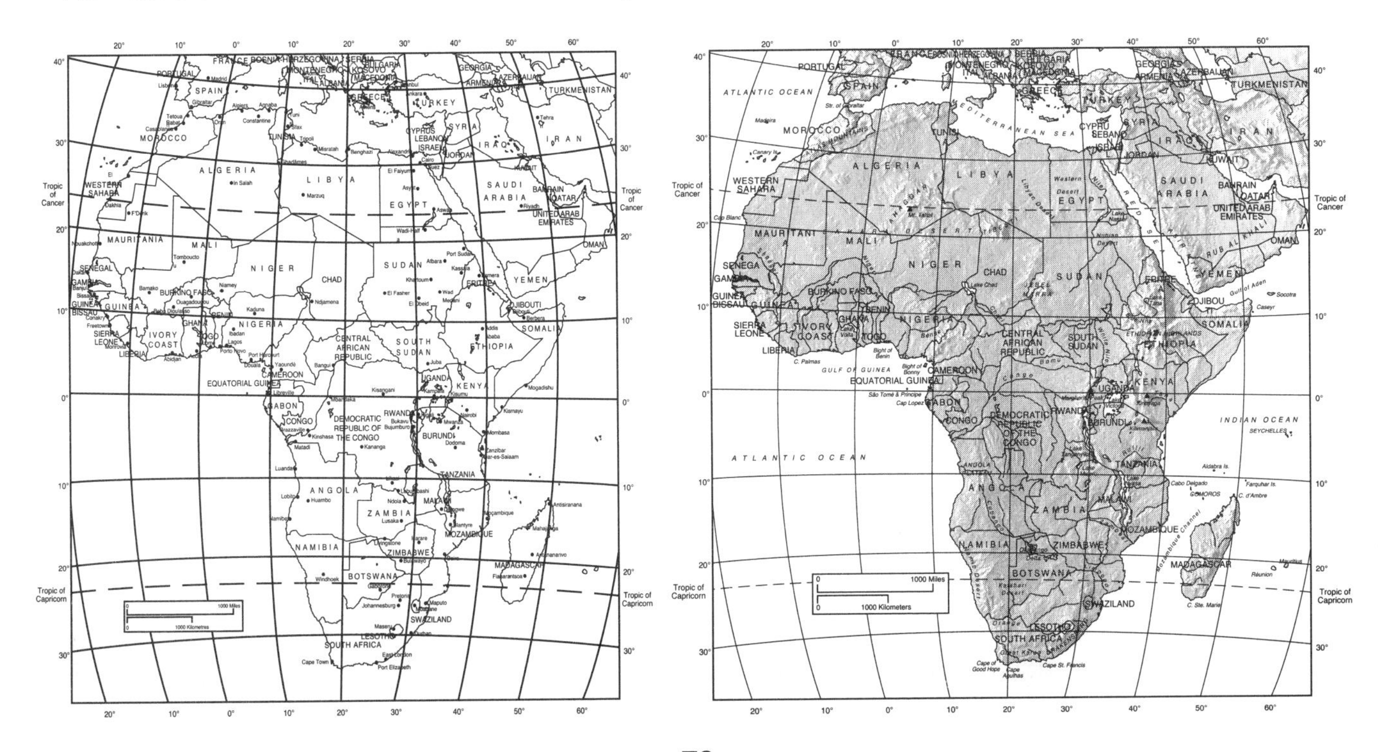

Name: ______________________________ Date: ____________________

Africa, South of the Sahara: *Location—Geoquest*

Safari, Anyone?

Directions: You are a tour guide in southern Africa. Your job is to create an advertisement for an African safari. Include relative and absolute location as well as an itinerary for the safari. Where will you go? The pilot will need the absolute location, or exact latitude and longitude coordinates, in order to get your guests to the right location to begin the trip. Write the advertisement below and include a map or pictures to catch the consumers' attention.

ADVERTISEMENT

____

____

____

____

____

____

____

____

Africa, South of the Sahara: *Place*

What makes southern Africa separate from northern Africa? What makes Africa, south of the Sahara, different from South America?

The answer lies in the second theme of geography, place. This allows people to have an accurate picture of why a place is unique. This theme of place can be defined by the physical and human characteristics that make a certain location different from other places on the earth's surface. The physical characteristics can be identified according to what is found in nature. The physical environment includes weather, climate, mountains, rivers, plants, animals, and natural disasters. These physical characteristics can and will change over time. For example, natural disasters or droughts can impact the physical environment. Since much of Africa, south of the Sahara, relies on subsistence farming, the people definitely are impacted by their physical surroundings. Tanzania, for instance, has to deal with little or no rainfall as they live on a subsistence basis. This means that the people grow their own food, make their own clothing and shelter, and have few goods left over to trade. The vast basin of the Congo River and the humid region around the Equator are covered in one of the world's last great rain forests. Madagascar is the island nation that lies in the Indian Ocean just east of Mozambique. The Cape of Good Hope is the southernmost point of the continent of Africa. Cape Town is an important port city for South Africa.

The human characteristics that impact a place are the man-made features, such as schools, buildings, roads, languages, religious practices, and governments. The African nations south of the Sahara have human characteristics that involve European influence. Much of this area was colonized by European nations. Therefore, the customs, religions, foods, and languages reflect this. For example, Zambia was a former British territory. Rwanda and Burundi formed part of German East Africa until after World War I. They were then placed under the League of Nations as the mandated territory of Ruanda-Urundi and were administered by Belgium. In 1962, they became two separate, independent states with a customs union between them. This union was ended in 1964, and each country has developed its own independent system.

Forests in Rwanda are cut down to create farmland or room for building.

The human characteristics of place can be positive or negative. The people of Rwanda have torn down valuable forests in order to clear the land for building and farming. Slowly, the negative effects of these decisions are being realized. However, governments are noticing the damage and are beginning to take action. The leaders are choosing to make wildlife and forest preserves. As you can see, human characteristics can change over time as well. Humans can make decisions that impact their place.

Name: ______________________________ Date: ______________________

Africa, South of the Sahara: *Place—Activity*

Name That Nation!

Directions: Using the following clue lists, identify the nation in Africa that is being described. The lists include physical and human characteristics to help you "Name That Nation!"

1. Famine; capital is Addis Ababa; Red Sea is the northeastern border; Great Rift Valley; the Blue Nile (Abay); droughts; exports are coffee, hides, oilseeds; Live Aid and Farm Aid to help people there

   Nation: ______________________

2. Lake Victoria in the north; Indian Ocean to the east; capital is Dodoma; includes the islands Pemba and Zanzibar; has the Selous Wildlife Reserve (one of the largest game reserves in the world); two-thirds of the land cannot be farmed due to the lack of water and swarms of disease-carrying tsetse flies; volcano of Ol Doinyo Lengai

   Nation: ______________________

3. Lakes make up eastern border; Congo River runs throughout; equatorial rain forests; the okapi, a forest animal related to the giraffe, is found only in this country; world leader in the production of copper, cobalt, and industrial diamonds; Pygmies; French is the official language

   Nation: ______________________

Using baskets to catch fish in the Congo River

4. Devastated by long years of civil war; Portuguese influence; capital is Luanda; diamonds and offshore oil; Atlantic Ocean to the west; mountains in the center; desert to the south; people grow sugar cane and cassava for a living

   Nation: ______________________

Name: ______________________________ Date: ____________________

Africa, South of the Sahara: *Place—Geoquest*

The Challenge of Apartheid

The policy of apartheid in South Africa is an example of place. It is a human characteristic that makes the nation of South Africa unique. Even though apartheid is no longer in effect, it is an important part of South Africa's history and human geography as well as part of American history. Hopefully, apartheid will be a policy that the human race can learn from so that history will not repeat itself.

Directions: Read the quote below. Research what apartheid was and what it meant for the people in South Africa. Then answer the questions below.

A black man identified only as "Lucky" told a foreign journalist, "You who are not even South African are freer to move around this country than I am. This is why we are frustrated here. Everywhere we turn we face restrictions and more restrictions."

1. Explain what this quote means to blacks in South Africa. ____________________

2. How did apartheid impact South Africa? ____________________

3. Why did apartheid become a racial policy in South Africa? ____________________

4. Explain the role of the United States with South Africa. ____________________

5. In your own words, define *apartheid.* ____________________

6. Describe who Nelson Mandela was and what influence he had in South Africa.

Africa, South of the Sahara: *Interactions*

Human-environment interactions, the third theme of geography, tells a very important story of how people must use, adapt, or change their environment in order to live. It also shows how the environment has impacted humans living there. In Africa, south of the Sahara, many interactions are taking place in order for people to survive. This part of the world has been dominated by foreign rule, which has left the continent with huge problems. Most of the people were poor and received very little education or training. The people of Africa, south of the Sahara, have had to struggle to unite differing groups of people and have had to fight off diseases, famine, droughts, and wars. The land in this area south of the Sahara includes thin pasture and scrub, giving way to **savanna** (grassland dotted with trees). There are some mountain ranges and the dusty deserts of the Kalahari and Namib. The humid regions around the equator and the vast basin of the Congo River are part of one of the world's last great rain forests.

Mining regions are extensive in sub-Saharan Africa; however, the mineral wealth is not evenly spread out across the region. Valuable minerals are found along the upper Atlantic coast and in countries south of the equator, such as South Africa, Guinea, and the Democratic Republic of the Congo. Fishing regions include the commercial fishing that is found off the southwestern coast of Africa. These regions are also home to the spiny lobsters that are trapped off the coast of South Africa.

Tragically, wildlife is under threat throughout the African continent. Many animals have been hunted to extinction. As humans build roads and cities, habitats are destroyed and migratory routes are blocked. Vital forests are cut down for timber and to clear land for farming. Without tree roots to anchor it, precious soil is washed or blown away in tropical storms. The land becomes too poor to grow anything, and it slowly turns into a desert floor. Humans can positively and negatively impact their environment. A positive measure taken by many governments is to try to conserve resources, create safe habitats, plant trees, and make parks. Solutions to negative interactions are not easy, and these conservation programs are expensive. In Africa, the interest of wildlife protection conflicts with the local people who need to graze their cattle or plant crops in order to make a living. Survival of humans is at stake in many of these African nations. However, many African nations now protect their wild animals in national parks. Some countries are also beginning to restock rivers and lakes with fish, and they are planting new forests. It is necessary to look at new ways of living with nature without destroying it.

The equatorial rain forest along the Lomami River in the Democratic Republic of the Congo

Name: ______________________________ Date: ______________________

Africa, South of the Sahara: *Interactions—Activity*

A Picture Is Worth a Thousand Words

Directions: Study the pictures from African countries located south of the Sahara Desert. Explain what you think is happening in each picture. Identify what type of interaction is taking place and determine whether it is positive or negative. Then look in a newspaper, a magazine, or the Internet for your own picture of an interaction to share with the class.

Guinea

____

Somalia

____

Kenya

____

Tanzania

____

Burundi

____

Democratic Republic of the Congo

____

Name: ______________________________ Date: ______________________

Africa, South of the Sahara: *Interactions—Geoquest*

Operation Preserve!

Directions: Many countries located in Africa, south of the Sahara Desert, are realizing that cutting down forests for farming and building has had negative consequences. The governments of these countries are trying to develop forest and wildlife preserves. They are, in some cases, even planting new trees. Rwanda is one country that is currently trying to make this human-environment interaction a positive one. Read the article on Rwanda. Then see how many countries in southern Africa are going through OPERATION PRESERVE or should begin these efforts in order to preserve precious forests and wildlife. Find out why the interaction is taking place and if the government and people are doing anything to promote change.

Rwanda to Reopen Gorilla Park

Rwanda's Parc National des Volcans (Volcanoes National Park), home to half of the world's rare mountain gorillas, will reopen next month for the first time since civil war forced its closure two years ago, park officials said Thursday. About 300 of the majestic mountain gorillas live in bamboo forests on the upper slopes of steep forested volcanoes inside the park, which lies on Rwanda's northern borders with Uganda and the Democratic Republic of the Congo.

Tourism was suspended at the park in June 1997, at the height of an armed insurgency waged by Hutu rebels who, in 1994, had led the genocidal slaughter of an estimated 800,000 Tutsis and moderate Hutus in Rwanda. But security has improved since a Rwandan-backed rebellion began in eastern Congo last August, depriving the Rwandan insurgents of vital rear bases across the border.

Another 300 mountain gorillas inhabit nearby Bwindi National Park inside Uganda, where eight foreign tourists were killed by Rwandan Hutu insurgents in March. Bwindi was closed immediately after the attack, in which a park ranger was also killed, but it reopened in April.

June 24, 1999
IWRC - Rehab News

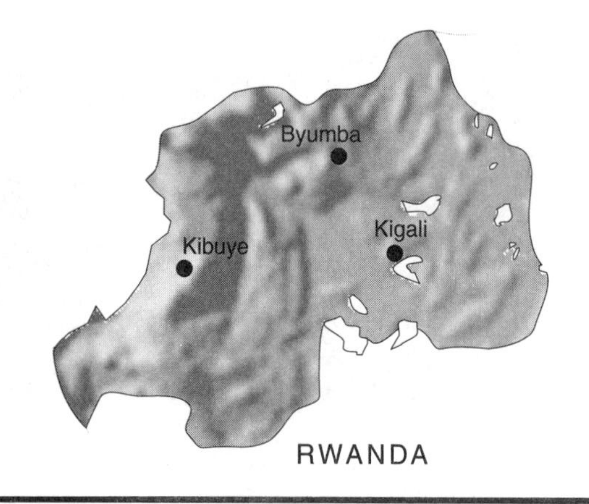

Africa, South of the Sahara: *Movement*

Movement, the fourth theme of geography, explains what people, ideas, and goods have moved. It indicates trade and diffusion of culture. Movement can also show how technologically advanced a particular culture is. For example, if one thinks about people moving, the questions that come to mind are "Where?" and "How?" Both of the answers will tell us if the people are moving near or far and by what means of transportation. In Africa, south of the Sahara, the people might use trucks or planes to move or perhaps a dugout boat or an ox-driven cart. The people may choose to move just down the Congo River or to the United States to attend a university. People move goods such as produce, livestock, or machinery throughout the continent.

An airport in Uganda

Many people believe that the Sahara Desert has negatively impacted movement between northern and southern Africa. But this is not accurate. The Sahara Desert did cut off Africans in the south from direct contact with Europe and western Arabia or the Middle East region. However, the Sahara Desert has always been a highway of trade and communications, especially along the Nile River.

One of the ideas that moved into southern Africa from the north was the technique of food growing and cattle raising. The knowledge of agriculture meant that larger populations could be sustained. Furthermore, only millet and sorghum are native to Africa. This means that other staple foods were moved into Africa from other continents. For example, rice, yams, and bananas were moved or imported to the east coast from Asia by traders. In the time of the slave trade, plants from the Americas, such as maize, cassava, and the sweet potato, were brought to West Africa to feed the slaves awaiting shipment. African farmers adopted these new foods and again added to how many people could be sustained.

Another example of movement would be the 1,000 different languages spoken in Africa by hundreds of ethnic groups. Arabic in the north, Swahili in the east, and Hausa in the west are the languages used by the largest numbers of people. In addition, because of European colonialism, English, French, and Portuguese are more widely used than any African language.

By 2008, political unrest in the region of Sudan had displaced millions of its citizens. Homes, farms, and businesses have been destroyed and numerous people have been turned into refugees. Desert refugee camps in Sudan and other countries of the region house tens of thousands of refugees. This mass movement has led to poor living conditions, overcrowding, and conflicts over food, water, and firewood.

Rivers provide a means of transportation in Africa, and many towns are located near rivers.

Name: ______________________________ Date: ____________________

Africa, South of the Sahara: *Movement—Activity*

Interdependence

Since many African countries south of the Sahara cannot produce everything they need, they must import and export. Nations are interdependent; they rely on each other for resources and goods.

Directions: Find out what the countries below import and export. Then answer the questions below. Some of these countries might have a long list of imports or exports. Choose the five major ones to record below.

COUNTRY	IMPORTS	EXPORTS
1. Angola		
2. Botswana		
3. Lesotho		
4. Madagascar		
5. Malawi		
6. Mozambique		
7. Namibia		
8. Swaziland		
9. Zambia		
10. Zimbabwe		

11. How does the United States play a role with Africa in interdependence?

____

____

12. Give five examples of trading partners that the above African nations depend upon.

____

____

Name: ______________________________ Date: ______________________

Africa, South of the Sahara: *Movement—Geoquest*

Languages on the Move!

Directions: Africa is known for the 1,000 languages that have been spoken throughout its history. If you traveled in southern Africa, would you be able to speak English and communicate with the people? Using the map of Africa below and the Internet or other research materials, create a map key and shade or label the major languages that are spoken in southern Africa.

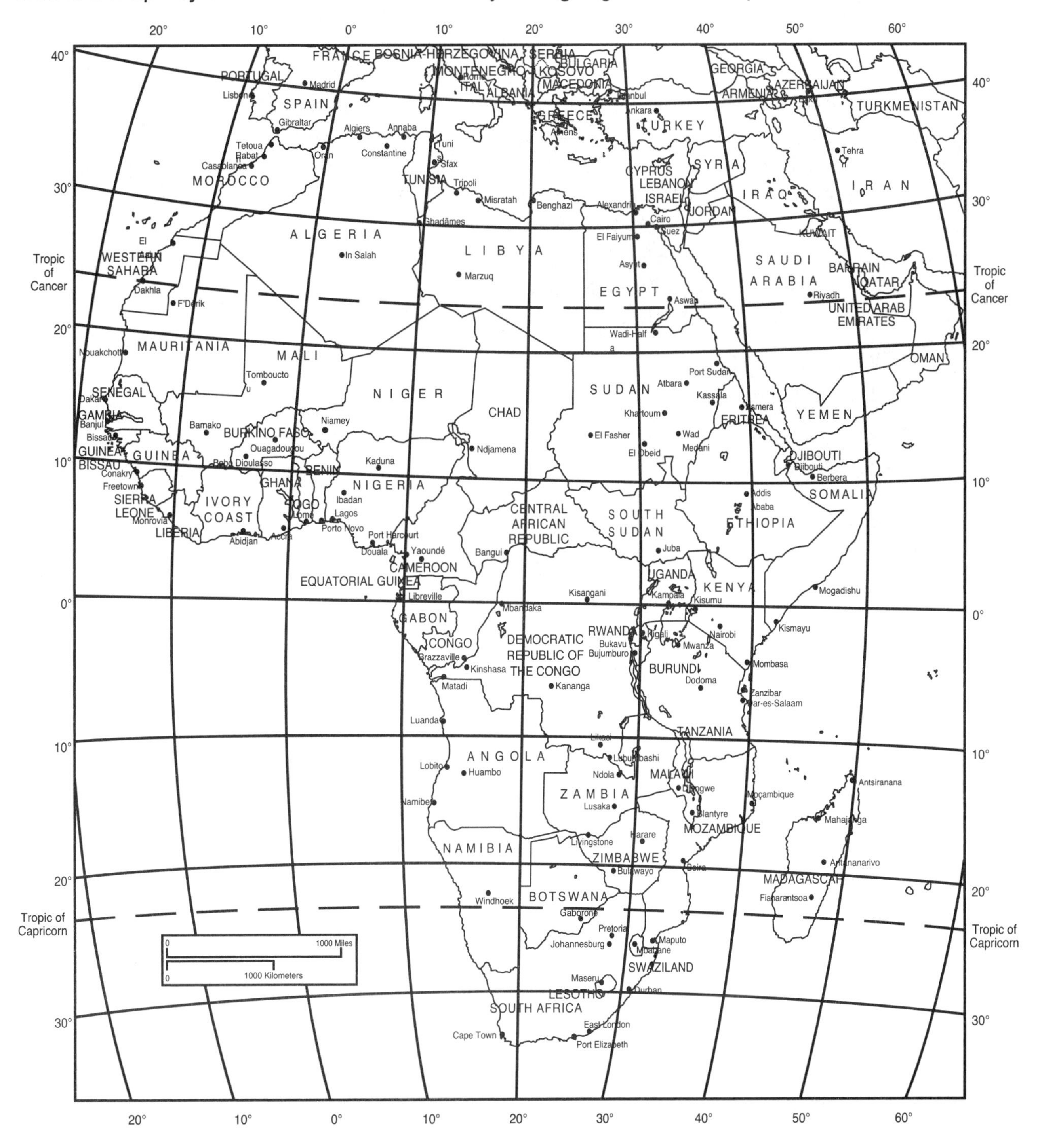

Africa, South of the Sahara: *Regions*

Geographers use the fifth theme of geography, regions, to better understand areas of the world. It is a way to analyze countries, cultures, and groups of people to determine which characteristics are shared and which are different. This has been more beneficial than simply studying individual countries. We live in a world that often looks negatively at being different. The theme of regions focuses on what people, cultures, or countries have in common with each other. However, it should not be just political borders, but also language, religion, government, or land characteristics that show the way regions are alike or different.

A diamond from the Kimberley Mine in South Africa

The entire world can be defined by regions, the physical and cultural characteristics that tie it together. These regions can be mapped. For example, one could map the rain forest regions of southern Africa. This area of the world could also be defined by its mining regions, its cotton regions, or its diamond regions. For example, the diamond regions are made up of the gem-producing countries South Africa, the Democratic Republic of the Congo, Liberia, Ghana, Tanzania, and Angola. Can you identify which are physical regions and which are cultural regions?

The Drakensberg Mountains

The study of regions also leads geographers to compare places by physical features. For example, one could look at the desert regions of southern Africa as compared to the Sahara Desert in northern Africa. Geographers can compare mountain regions such as the Drakensberg Mountains and the Ruwenzori Mountains in southern and central Africa to the Pamir Mountains and the Himalaya Mountains in Asia. Even though these are all mountain regions, they are different due to the climate, location, and influence of humans. The Drakensberg Mountains are located along the coast of the Indian Ocean and experience a milder climate compared to the Himalayas, which are in inland Asia.

Language regions tend to get complicated because there are so many dialects and tribal languages in southern Africa. It makes their culture interesting and unique, but more difficult to map.

Regions can change over time due to weather and climate patterns, economic conditions, conflict, accessibility to trade routes, technology, and many more factors. Geographers have to keep studying regions in order to predict the future needs of people or environmental changes. For instance, in Rwanda, if people continue to tear down forests for agriculture and building, their forest region will change. This will also impact the wildlife regions and the agricultural regions. The study of changing regions is essential for the future of our cultures and our environments.

Africa, South of the Sahara: *Regions—Activity*

You Determine the Regions!

Directions: Read the information boxes below. They show characteristics of several African countries located south of the Sahara Desert. Answer the following questions on your own paper.

Angola	Namibia	Tanzania	Mozambique
Area: 481,354 sq. mi. **Population:** 19,088,106 **Capital:** Luanda (5,068,000) **Other major cities:** Huambo (979,000), Lobito (805,000), Benguela (128,000) **Highest point:** Moco (8,593 ft.) **Official language:** Portuguese **Main religions:** Christianity, traditional beliefs **Currency:** Kwanza **Main exports:** Crude oil, coffee, diamonds, fish products, sisal, corn, palm oil **Government:** Multiparty republic **Per capita GDP:** 2013 Est. U.S. $6,300	**Area:** 318,250 sq. mi. **Population:** 2,198,406 **Capital:** Windhoek (342,000) **Other major cities:** Swakopmund (45,000), Rundu (64,000), Rehoboth (29,000) **Highest point:** Brandberg (8,462 ft.) **Official language:** English **Main religions:** Christianity, traditional beliefs **Currency:** Namibian dollar **Main exports:** Diamonds, uranium, fish products, meat products, livestock **Government:** Multiparty republic **Per capita GDP:** 2013 Est. U.S. $8,200	**Area:** 364,879 sq. mi. **Population:** 49,639,138 **Capital:** Dodoma (411,000) **Other major cities:** Dar-es-Salaam (3,207,000), Mwanza (707,000) **Highest point:** Mt. Kilimanjaro (19,340 ft.) **Official languages:** Swahili, English **Main religions:** Islam, Christianity, traditional beliefs **Currency:** Tanzanian shilling **Main exports:** Coffee, cotton, sisal, cloves, diamonds, salt, cement **Government:** Multiparty republic **Per capita GDP:** 2013 Est. U.S. $1,700	**Area:** 308,640 sq. mi. **Population:** 24,692,144 **Capital:** Maputo (1,589,000) **Other major cities:** Matola (761,000), Nampula (472,000) **Highest point:** Mt. Binga (7,990 ft.) **Official language:** Portuguese **Main religions:** Traditional beliefs, Christianity, Islam **Currency:** Metical **Main exports:** Cashew nuts, sugar, cassava, petroleum products, aluminum, copra, cotton **Government:** Multiparty republic **Per capita GDP:** 2013 Est. U.S. $1,200

1. Identify what type of characteristics are shown in the information boxes.
2. The countries of Mozambique and Tanzania border each other. Compare and contrast their characteristics. What do they share in common? What are their differences?
3. The countries of Angola and Namibia border each other. Compare and contrast their characteristics. What do they share in common? What are their differences?
4. How can geographers use the above information? Using a map of Africa, draw or shade in the regions.
5. Study the export list for each of the four countries. What conclusions can you make about the environment? How does the physical environment impact the culture of these nations?

Name: ______________________________ Date: ____________________

Africa, South of the Sahara: *Regions—Geoquest*

You Be the Geographer!

Directions: There are two types of regions below. Choose one. Study each picture. Write down the importance of the region being addressed. Then research where else in Africa or in the world you might find this type of region. Shade these regions on a map of Africa. Lastly, compare or contrast these regions to the area in which you live. Are there any of these regions in your state or in the United States? Explain.

Mining Regions

Animal Preserve

Asia, Oceania, Australia, and Antarctica: *Location*

More than half of the world's population live in the area of the world known as the Asian region. Over three billion people call this area, which covers over 17,400,000 square miles, home.

At one time this region was known as the Far East. In the early years of trade and exploration, European merchants traveled the globe looking for goods to buy and sell. In order to get to this area, they would have to travel far to the east. When looking at the Asian region from the perspective of North America, is the Far East really "far east"?

To answer that question, we need to turn our attention to the first theme of geography—location. The **absolute location** of the Asian region can be found by examining the coordinates of latitude and longitude on a map for a specific area. The absloute location of Asia is roughly 50 degrees North latitude to the equator and 60 to 160 degrees East longitude. The **relative location** of the region can be identified as east of the Ural Mountians, south of the Arctic Circle, west of the Pacific Ocean, and north of the Indian Ocean.

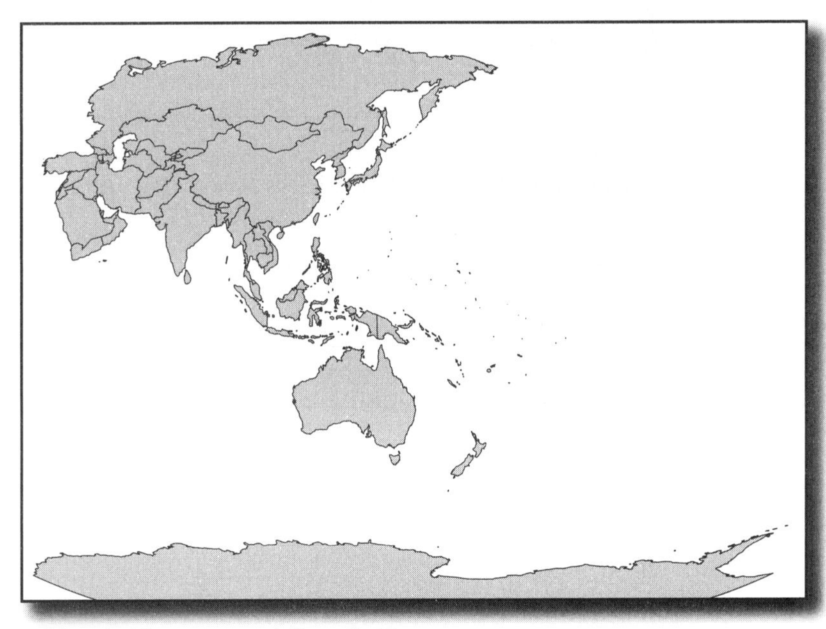

The Asian region can be further subdivided into four major areas: Southwest Asia (the Middle East), the Indian Subcontinent, Southeast Asia, and Eastern Asia. We have already discussed the countries of the Middle East, such as Saudi Arabia, Israel, Iraq, and Afghanistan. Countries located on the Indian Subcontinent include India, Bhutan, Pakistan, Nepal, and Bangladesh. Located in Southeast Asia are countries such as Indonesia, the Philippines, Thailand, Vietnam, and Singapore. Eastern Asia includes countries like China, Mongolia, Russia, Japan, North Korea, and South Korea.

In this unit, we will also be discussing the continent of Australia, the islands of Oceania, and the continent of Antarctica.

The location of these countries and their cities vary greatly due to the vastness of the region. Terrain ranges from mountainous to tropical. A great variety of products are also produced, including fruits and vegetables, metals and gemstones, and livestock.

Name: ______________________________ Date: ____________________

Asia, Oceania, Australia, and Antarctica: *Location—Activity*

Directions: Using the reading and an almanac, answer the following questions.

1. List the four subdivisions of the Asian region.

   a. ____________________

   b. ____________________

   c. ____________________

   d. ____________________

2. Why was part of the region also called the "Far East"? ____________________

   ____

   ____

3. Identify the absolute location of the following Asian cities.

   a. Phnom Penh, Cambodia ____________________

   b. Jakarta, Indonesia ____________________

   c. Hanoi, Vietnam ____________________

   d. Manila, Philippines ____________________

   e. Beijing, China ____________________

   f. New Delhi, India ____________________

4. How far (in miles) is the city of Chicago, Illinois, from the following Asian cities?

   a. Bangkok, Thailand ____________________

   b. Kuala Lumpur, Malaysia ____________________

   c. Tokyo, Japan ____________________

   d. Katmandu, Nepal ____________________

Name: ______________________________ Date: ______________________

Asia, Oceania, Australia, and Antarctica: *Location—Geoquest*

Directions: Using the space provided, brainstorm a list of ideas that you have about Asia. Include location, distance from the United States, population, climate, products, government, and so on. Compare your list with a classmate's. Together, present your joint findings in a multimedia presentation to the class.

Asia, Oceania, Australia, and Antarctica: *Place*

When looking at the second theme of geography, place, one needs to look at the special characteristics that make that place unique on the earth. Asia, Oceania, Australia, and Antarctica—each has different physical characteristics and human characteristics.

The Yellow (Huang) River in China

Physical characteristics of the Asian region include a vast natural environment with tropical jungles, wetlands, and large rivers. Some of these rivers include the Yellow River (or Huang River), Mekong River, Amur River, the Brahmaputra River, and the Ganges River. When looking at Central Asia and the Indian Subcontinent, there are mountain ranges. The highest point in Central Asia is Mount Everest on the Tibet-Nepal border in China with a height of 29,029 feet (8,848 meters). The climate changes with each location. In the mountain regions, temperatures can get many degrees below freezing. Near the tropical coastlines, temperatures will be more temperate. Monsoons (periods of rainy, stormy weather) affect life in the more tropical areas. Australia's natural environment consists of tropical coastal regions with large deserts inland. Oceania is made up of island countries such as French Polynesia, Fiji, Micronesia, and New Zealand. Antarctica is primarily a frozen, barren land. It includes mountains and ice shelves.

The Pacific island nation of Fiji

China's emperors once lived in the Forbidden City in Beijing.

Human characteristics are equally as diverse. Systems of government in Southeast Asia have typically been monarchies. A **monarchy** is a form of government where the people are ruled by a king, queen, emperor, or empress. The monarch is usually believed to be descended from the gods. For centuries, emperors ruled over China and Japan. Language is also extremely diverse. Hundreds of different languages and variations are spoken across the continent. Human dwellings were built to adapt to the surrounding terrain and climate. Houses on stilts and pilings are typical along the oceans and seas where flash floods are common. However, many major cities have large modern high-rises and big businesses.

Name: ____________________ Date: ____________________

Asia, Oceania, Australia, and Antarctica: *Place—Questions*

Directions: Answer the following questions.

1. Describe the physical characteristics of the following countries.

   a. China ____________________

   b. India ____________________

   c. Japan ____________________

   d. Australia ____________________

   e. Philippines ____________________

2. Describe the human characteristics of the following countries.

   a. China ____________________

   b. India ____________________

   c. Japan ____________________

   d. Australia ____________________

   e. Philippines ____________________

3. Define:

   a. Monsoon ____________________

   b. Monarchy ____________________

Name: ________________________ Date: ________________

Asia, Oceania, Australia, and Antarctica: *Place—Geoquest*

Directions: Choose one or more of the following activities to complete. You may use any variety of traditional and/or digital media technology available to demonstrate your knowledge.

1. Select different archipelagoes in the Asian region and prepare maps showing the names of each of the islands in the group.

2. Select a river in the one of the regions discussed in this unit and prepare a travelogue about the river. Describe the journey and what you might see and do.

3. Prepare a display on Mount Everest, showing where it is, who lives in the surrounding area, plants and animals in the surrounding area, and famous climbing expeditions and their dates.

4. Select a desert in Australia. Prepare a poster or display showing the types of vegetation and animals that live there.

5. Research a volcano in the Philippines or Indonesia. Create a time line with pictures and information describing the activity of the volcano over the years.

Mt. Everest in the Himalayas

Asia, Oceania, Australia, and Antarctica: *Interactions*

Human-environment interaction is the third theme of geography. How do humans affect the environment and how does the environment affect humans in the Asian region?

Humans in this region have long used the natural environment for farming and agriculture. Rice paddies dot the countryside all along the coastal areas. Farmers learn to read the weather to know when the approaching wet season, or **monsoon season**, is coming. Rivers and waterways are used for farming and transportation. People also dam up rivers to use the water for farming.

Flooded rice fields during monsoon season

In addition, there has been a long history of conflict among humans in the Asian region. Conflicts have been a part of life. Several countries have undergone civil wars, including China in 1949, Vietnam in the 1960s, and Kampuchea (Cambodia) in the 1970s. Countries such as India, Pakistan, China, and Vietnam have also challenged each other over borders.

An explosion in the crater of the Lokon-Empung dual volcano in Indonesia is visible from the town only five kilometers away.

Nature, in the form of volcanoes and monsoons, has affected human life as well. While volcanoes can be destructive and have forced humans to leave certain areas, they have also created many of the islands in the region.

The Great Barrier Reef is known for its natural sea life. It brings many tourists to Australia, positively affecting the economy. However, humans who disturb the reef and its wildlife have a negative impact on the reef.

The quality of life for the people of the Asian region is not very high. Villages often have little or no pure water supplies. Families live in crowded conditions that foster disease and poor health. The average life expectancy is approximately 54 years.

Religion in Asia is as diverse as its people. Some of the major religions include Hinduism, Islam, Buddhism, Confucianism, and Taoism. Each religion is vastly different from another. Since the early days of Christianity, missionaries have made their way east into India, central Asia, and the Far East to make converts and establish churches. Priests and missionaries often traveled with the early explorers and traders. An offshoot of this work was the spread of Western culture and influence into the Asian region. Likewise, Asian culture made its way to the Western world.

Name: ______________________ Date: ______________________

Asia, Oceania, Australia, and Antarctica: *Interactions—Activities*

Directions: The following might have been headlines in a newspaper during Asia's history. Do some research, and write a sentence explaining each headline's meaning.

1. Civil War Erupts in Vietnam ______________________

2. Earthquake and Tsunami Devastates Japan ______________________

3. Chinese Culture Invades the Islands ______________________

4. Emperor Makes Declaration ______________________

5. Missionaries Forced to Leave China ______________________

6. Japan Loses All ______________________

7. Rainy Season Produces Record Flooding ______________________

8. Roald Amundsen Reaches South Pole ______________________

Name: ______________________ Date: ______________________

Asia, Oceania, Australia, and Antarctica: *Interactions—Geoquest*

Directions: Choose one or more of the following activities to complete. You may use any variety of traditional and/or digital media technology available to demonstrate your knowledge.

1. Compare and contrast the different religions and philosophies of Asia: Hinduism, Islam, Buddhism, and Confucianism. Create a chart.

2. Research the caste system of India. Prepare a diagram of the major castes and explain how the system works.

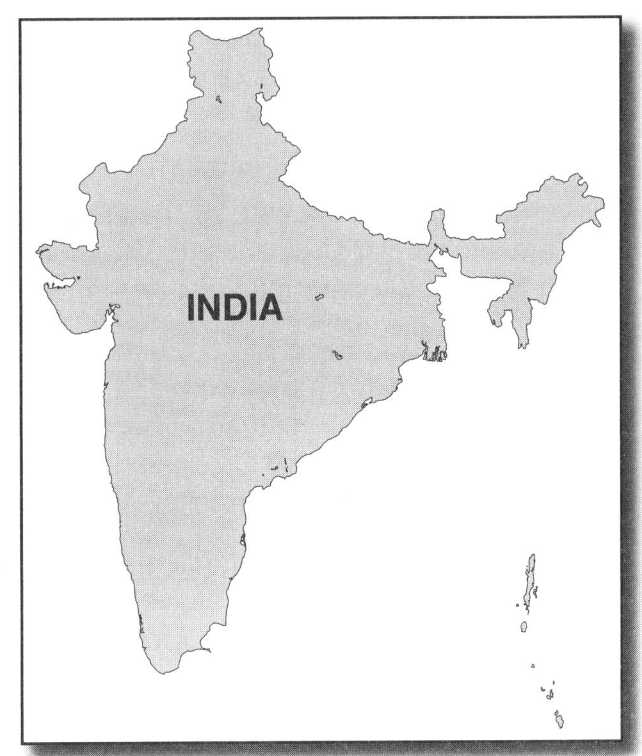

3. Find out where rice is grown on continents throughout the world. Create a map of the rice-growing regions.

4. Design a model of the summer and winter monsoons. Gather news articles on this year's monsoons and begin a monsoon news watch.

5. Research where dangerous tsunamis have occurred in the past 20 years. Map these areas on a world map. List the number of lives lost in particular tsunamis.

Asia, Oceania, Australia, and Antarctica: *Movement*

The movement of goods, services, and ideas around the globe is the fourth theme of geography. The Asian region has been greatly influenced by the spread of Western religion and culture.

Before the age of exploration, the Far East was largely unexplored. At one point, the ancient Chinese built the Great Wall of China to keep out nomadic herders and invaders. Only traders, such as Marco Polo from Italy, made the long, overland trek to the Orient. European traders brought their goods and ideas to Asia, and then took Asian goods and ideas back to Europe.

When Christopher Columbus set off on his famous voyage, it was not to find a new world, but to find a quicker, easier route to Asia. Although he was not successful, other explorers followed his example. In the following centuries, explorers came to find the many rich lands of the Asian region, bringing to Asia Western culture and religion. At the same time, Asian influence spread to Europe and the Americas.

With the end of World War II in 1945, much of Asia had suffered the destruction of war. Japan was heavily damaged; yet, by the 1980s, it had been rebuilt and was showing itself to be a technological giant. Japanese-built automobiles, computers, televisions, and other devices began to flood Western markets. Japan was able to move its goods and services around the globe.

Australia was originally created as a prison colony for England. Prisoners were forced to relocate to Australia. Eventually it became a new frontier where settlers came voluntarily. The Polynesian peoples moved throughout the islands of Oceania in various types of boats and rafts constructed from native materials.

The exploration of Antarctica and the race to become the first to reach the South Pole was a popular endeavor in the early twentieth century. The Norwegian explorer Roald Amundsen reached the South Pole on December 16, 1911. He was followed closely by Robert F. Scott on January 14, 1912. Today the only human inhabitants of Antarctica are scientists who live in various research bases and study the polar region.

With the onset of the Internet and World Wide Web, ideas are now able to be transmitted across the globe in a matter of seconds. No longer is there a need to wait days or weeks for mail delivery from the other side of the world. Asian countries, such as South Korea, China, Japan, Taiwan, India, and Vietnam, are leaders in producing the electronic devices used in the Information Age. Many of their citizens are able to take advantage of these devices and services.

Hong Kong is a modern city connected to the rest of the world through trade, communication, and business.

Name: ______________________________ Date: ______________________

Asia, Oceania, Australia, and Antarctica: *Movement—Activity*

Directions: Answer the following questions.

1. Define *movement* and give an example using this region. ______________________

____

____

2. Who was Marco Polo, and what impact did he have on the region? ______________

____

____

____

3. Describe the changes that occurred in Japan after 1945. ______________________

____

____

____

4. What impact does modern technology have on you? Explain. __________________

____

____

____

5. Explain the purpose of the Great Wall of China. ____________________________

____

6. Who were the first Europeans to live in Australia, and why were they selected to go there?

____

____

7. Who was the first person to reach the South Pole, and why was this significant?

____

____

____

Name: ______________________________ Date: ____________________

Asia, Oceania, Australia, and Antarctica: *Movement—Geoquest*

Directions: Choose one or more of the following activities to complete. You may use any variety of traditional and/or digital media technology available to demonstrate your knowledge.

1. Prepare a time line of the history of China, Japan, or India.

2. Prepare a report on trade between the United States and Japan.

3. Research several explorers and note their contributions to the spread of Western culture to Asia.

4. Prepare a map of Antarctica. Research the race to the South Pole, and draw the routes taken by Amundsen and Scott to the South Pole.

Roald Amundsen's polar expedition

5. Make a list of items in your home or school that have been made in an Asian country.

6. Trace the origin of one of your electronic devices. Use a map to show where it was made and a likely route from that country to your home or school.

7. Make a list of words, traditions, or items from Asia that have become common parts of Western civilization.

Asia, Oceania, Australia, and Antarctica: *Regions*

The fifth theme of geography is regions. Within the Asian region are many different, smaller regions. The Asian region can be subdivided into Southwest Asia, Southeast Asia, Eastern Asia, and the Indian Subcontinent. Oceania and the continents of Australia and Antarctica are also included in this unit.

The Southeast Asian region consists of Indonesia, the Philippines, Myanmar (Burma), Thailand, Malaysia, Singapore, Cambodia, Laos, and Vietnam. These countries are primarily tropical and depend on the seas and oceans for much of their lifestyles. The religions and languages in each of these countries vary greatly depending on your specific location. Government and cultures are also vastly different.

The Eastern Asian region is made up of China, Mongolia, North Korea, South Korea, Japan, and Taiwan. The eastern part of Russia is also in this region. The northern areas of these countries are mountainous while the southern areas are more coastal. Again, religion, language, government, and culture are very different in each location.

On the Indian Subcontinent are India, Bhutan, Pakistan, Nepal, Bangladesh, Maldives, and Sri Lanka. These countries have three seasons: cool from October to February, hot and dry from March to June, and rainy from July to September. Using this cycle, farmers are able to plan the crops they will plant and harvest.

Oceania, Australia, and Antarctica each have their own individual cultures, governments, religions, and languages.

Oceania is made up primarily of many small island countries located in the South Pacific. Included in this region are the Solomon Islands, New Caledonia, Micronesia, Kiribati, Fiji, Tonga, Western Samoa, American Samoa, the Cook Islands, and French Polynesia. Different native languages are spoken here as well as the languages of the European nations who colonized the islands at one time or another.

Australia is a country and continent in itself. English is the official language of Australia and neighboring New Zealand, as both were colonized by the British. The Great Dividing Range of mountains runs along the eastern part of Australia. Much of the Western Plateau of the country or "outback" is desert. In the grasslands of the Central Lowlands, farm animals such as sheep and cattle are raised.

Antarctica is a frozen, barren continent surrounding the South Pole. The Transantarctic Mountains cross the continent, and ice shelves extend out from the coast into the surrounding seas. Several species of wildlife, such as the emperor penguin and the southern elephant seal, live in Antarctica. However, the only humans who live there are scientists from other countries who live on scientific research bases.

Name: ______________________________ Date: ____________________

Asia, Oceania, Australia, and Antarctica: *Regions—Activity*

Directions: Research these countries and fill in the following information on the chart.

Country	Form Of Government	Leader	How Chosen	How Long in Power	Political Parties
Myanmar (Burma)					
Australia					
Vietnam					
New Guinea					
Thailand					
Singapore					
India					
China					
Japan					
Philippines					

Name: ______________________________ Date: ______________________

Asia, Oceania, Australia, and Antarctica: *Regions—Geoquest*

Directions: Choose one or more of the following activities to complete. You may use any variety of traditional and/or digital media technology available to demonstrate your knowledge.

1. Prepare a travel itinerary for a trip around the Asian region.
2. Find websites with photographs of different Asian countries and tourist sites. Prepare a computer-aided presentation.
3. Prepare a series of maps showing Asia now, in this year, and 100 years ago. Note major changes.
4. Select one aspect of Asian culture and prepare a report for the class.
5. Prepare an Asian food and share it and the recipe with the class.
6. Locate and bring into class some products from the Asian region that are available in the United States.
7. Find websites or magazines of different Australian tourist sites, landforms, and wildlife. Prepare a presentation for the class.
8. Research Antarctica. Prepare a map showing which countries claim territory in Antarctica. Identify and locate scientific research bases on the map.

Uluru (Ayers Rock) in Central Australia is considered a sacred place by the Aborigine people.

9. Research one of the island nations of Oceania. Prepare a travel brochure for the country.
10. Research the Indian Ocean Tsunami of 2004. What advanced warning systems were in place? What newer systems are in place today? Has much changed? Why or why not?

Name: ______________________________ Date: ______________________

North America

Name: ______________________ Date: ______________________

South America

Name: ______________________________ Date: ______________________

Europe

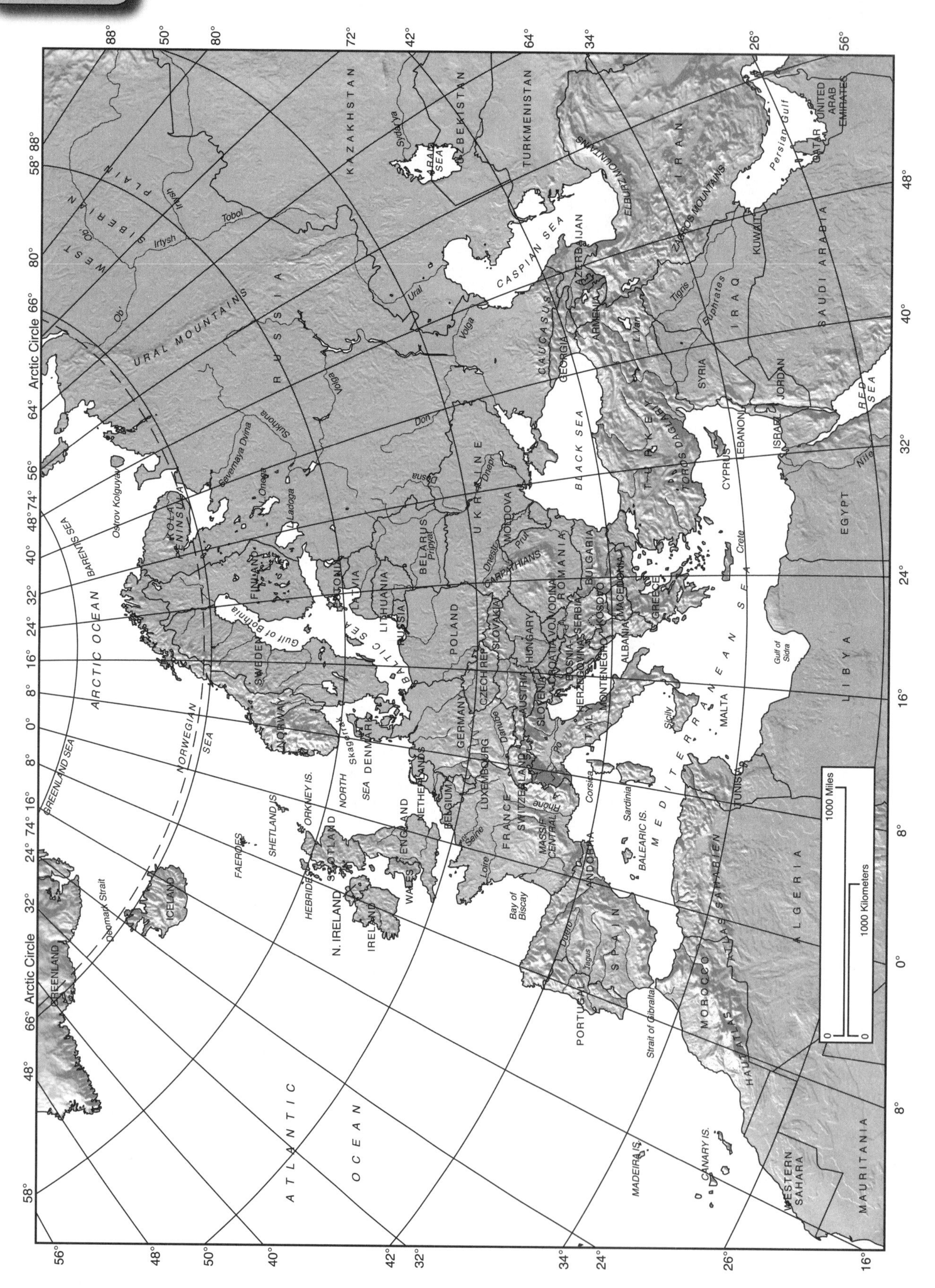

Name: ______________________________ Date: ____________________

Asia

Name: ______________________ Date: ______________________

Africa

Name: ______________________________ Date: ______________________

Australia/Oceania

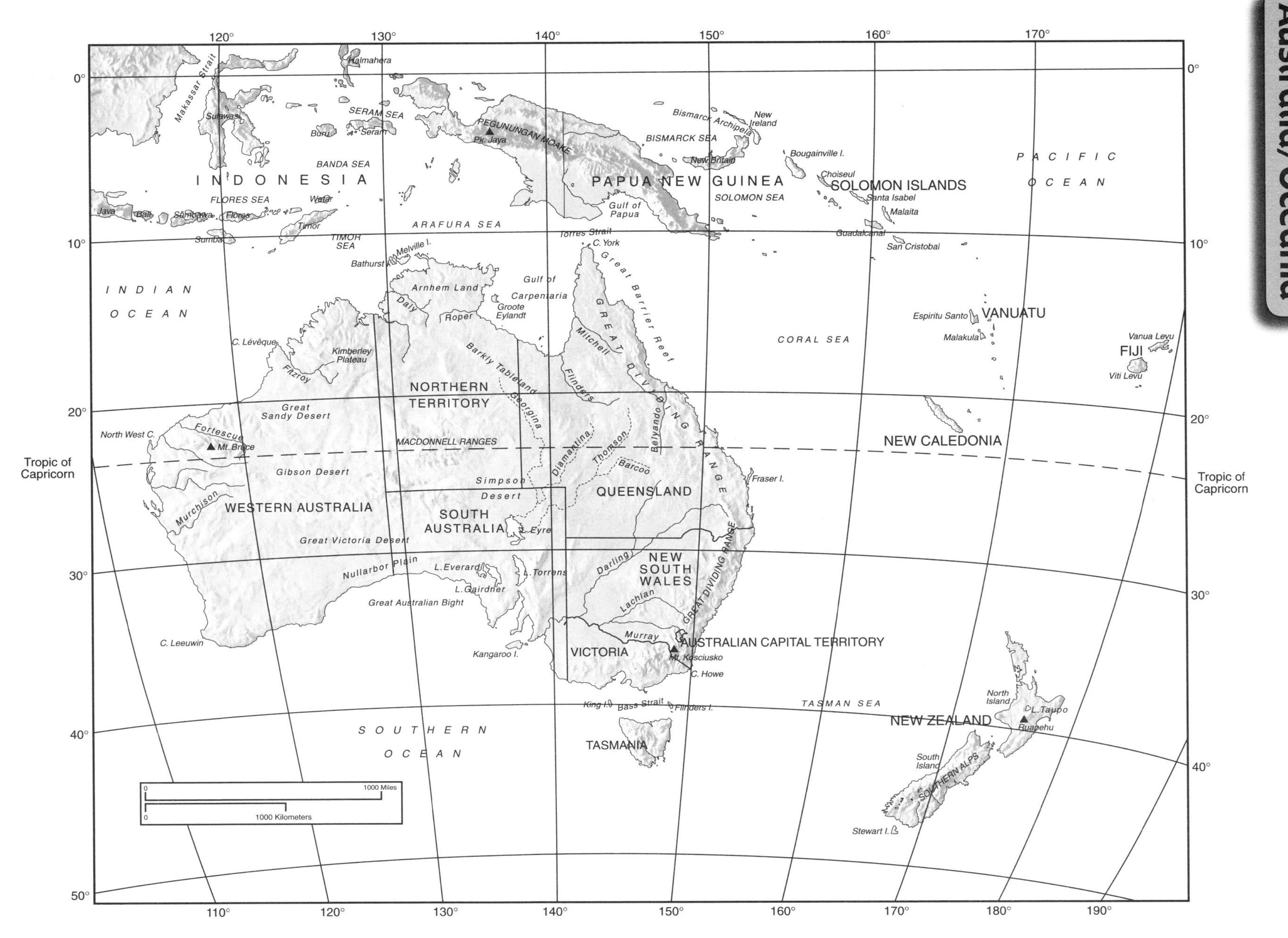

Name: ______________________________ Date: ______________________

Antarctica

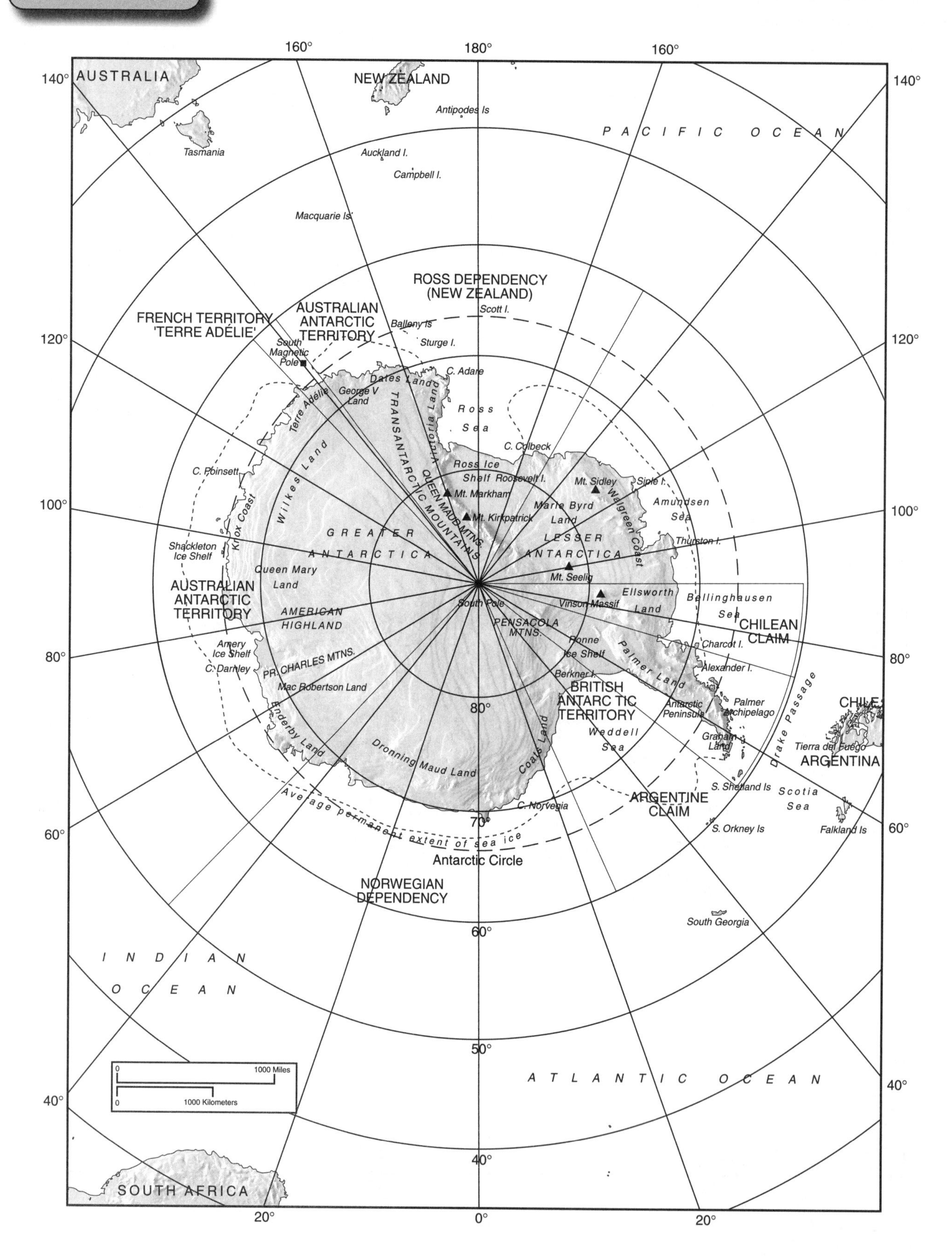

Name: ______________________________ Date: ____________________

World Map

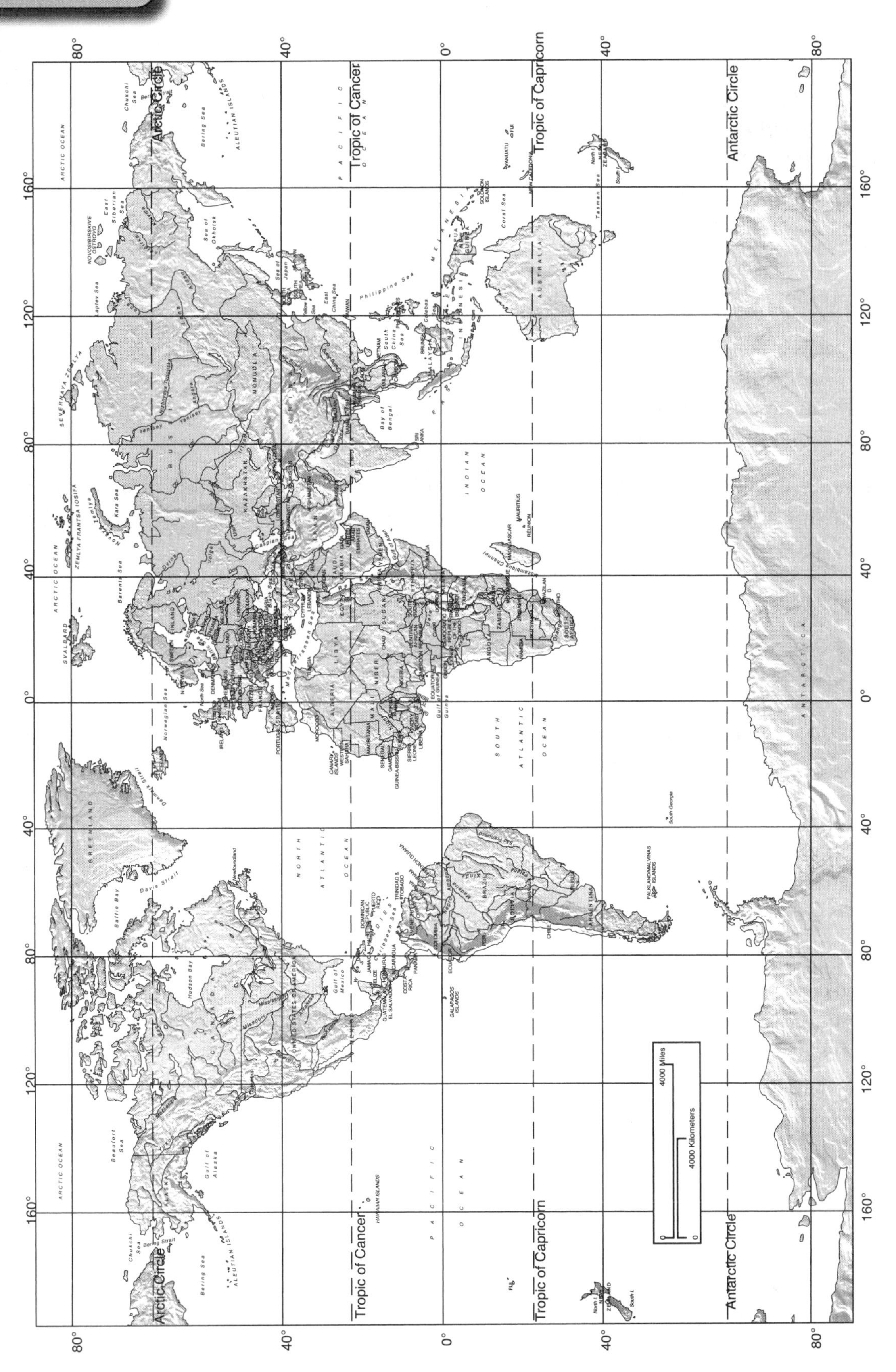

Global Summit Activity

Teacher Note: This is a project called Global Summit. There are two parts: the first part focuses on the five themes of geography, and the second part focuses on foreign policy. It is designed to be an individual project that develops higher-level thinking. Encourage your students to choose from any of the wide variety of current multimedia Apps such as Prezi™, i-Movie™, Keynote™, Explain Everything™, Google Presentation™, etc., to demonstrate their knowledge regarding their Global Summit activity.

Part 1:

1. The student will randomly pick a country. Look for the country in the news. The student will be looking at a country through the lens of the five themes of geography and current events.
2. The student will begin to investigate his or her country and determine what the conditions are for a child who lives in that country.
3. The student will study the geography of a country using the five themes listed below. The student will take his or her findings and create a scrapbook. **THE GLOBAL SUMMIT SCRAPBOOK SHOULD INCLUDE:**
   a. **Location:** Relative—Use a world map, shade in your country, and write out its relative location. Make a country map showing major rivers, cities, the capital, mountains, and deserts. Absolute—In writing, describe the country's absolute location by finding out its exact latitude and longitude location.
   b. **Place:** Physical Characteristics of Place—Find three pictures of your country that show its physical characteristics. Write a brief paragraph under each picture describing your country's physical characteristics. This should include bodies of water, deserts, mountains, climate, and weather conditions. Human Characteristics of Place—Find three pictures from your country that give evidence of human characteristics (man-made features). Write a brief paragraph under each picture describing your country's human characteristics.
   c. **Human-Environment Interactions:** Find out what type of interactions are going on in your country between the people, animals, land, and the environment. These could be positive or negative. Draw or find one picture that is an example of interactions in your country. Write two or three paragraphs that explain interactions in your country. Why are these interactions necessary? How have the interactions changed throughout your country's history?
   d. **Movement:** What is being moved in your country? Have ideas or religions moved? Have borders moved in your country? Have there been any wars or conflicts to impact movement? Identify what raw materials, natural resources, or finished products are being moved. Each student should make a product map showing the movement of goods and people.

Global Summit Activity (cont.)

e. **Regions:** In what region of the world does your country belong? Show this region as another color on your world map. Explain if your country's region is developing or developed and provide evidence of this. Think about this: Would it be a safe region of the world for a child? Why or why not? Finally, within the country you selected, are there various regions? Explain. For example, in the United States, we have a corn belt, a wine region, and different industrial and agricultural regions.

*** Add any current events you find for your country.

*** Add a conclusion page in which you analyze the geography of your country, determine your country's capability to sustain life, and the economic situation for children. This is your opinion based on the findings above.

In summary, there should be five parts to your scrapbook plus a current events section (this depends on the happenings in your country) and a conclusion page.

4. Next, the students will be writing (in class) a letter to their country's embassy to gather information to use for Part 2 of this project. One website for embassy addresses is <www.state.gov/s/cpr/rls/dpl/32122.htm>. This is the website for the U.S. Department of State.

Part 2:

1. The student will continue with the research of the same country; however, the focus will be foreign policy, UN and NATO participation, and the state of the children. The student will be making some decisions on the future outlook of his or her country, and he or she will write his or her own United Nations Resolution. He or she must back up the conclusions with data. The information and data can be found at UNICEF's website <www.unicef.org>. UNICEF has done extensive research annually on the State of the World's Children.
2. In addition, the student will be determining where the country belongs on Maslow's Hierarchy of Needs. (See the diagram on the next page.)
3. Finally, the student will take his or her information, research, and conclusions and create a three- to five-minute multimedia presentation. The presentation should include the following sections.
   a. Overview of the five themes of geography (using scrapbook information)
   b. Foreign policy of the country—include UN and NATO participation
   c. Summary of the State of the Children—information from <www.unicef.org>
   d. Determine the country's current status on a hierarchy of needs chart (see example on page 118)
   e. Current events and future outlook summary
   f. The proposal of a UN resolution to provide a solution for one of your country's problems

Global Summit Activity (cont.)

Maslow's Hierarchy of Needs:

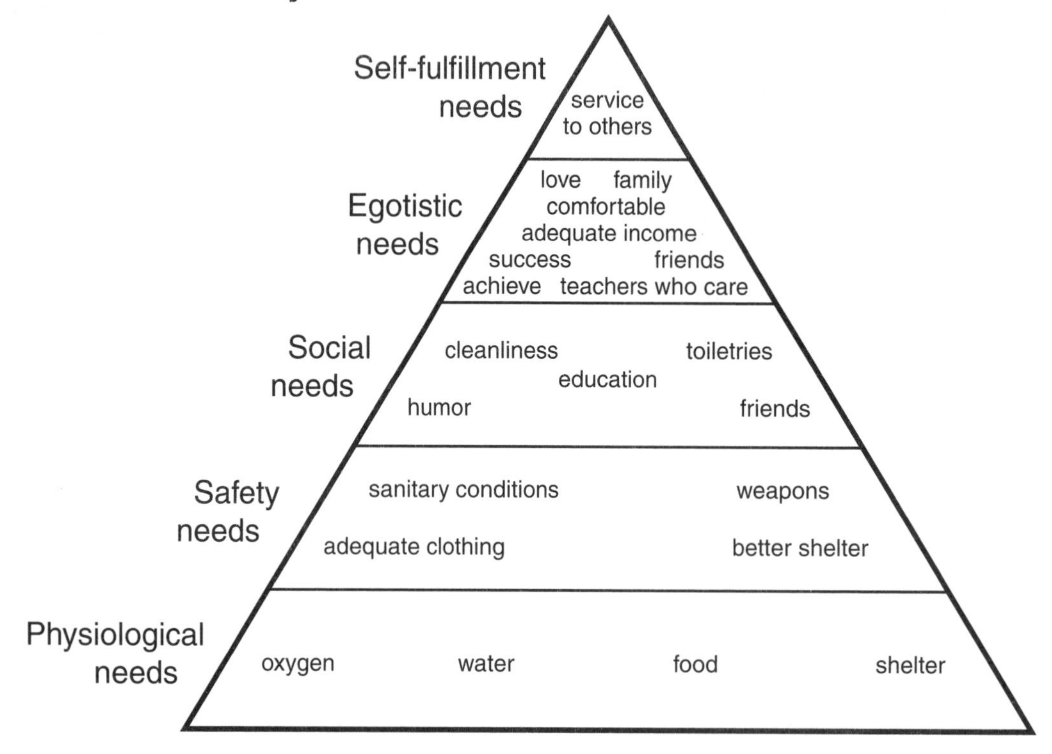

Example of rating a country on Maslow's Hierarchy of Needs:

(What level of needs are currently being met by that country?)

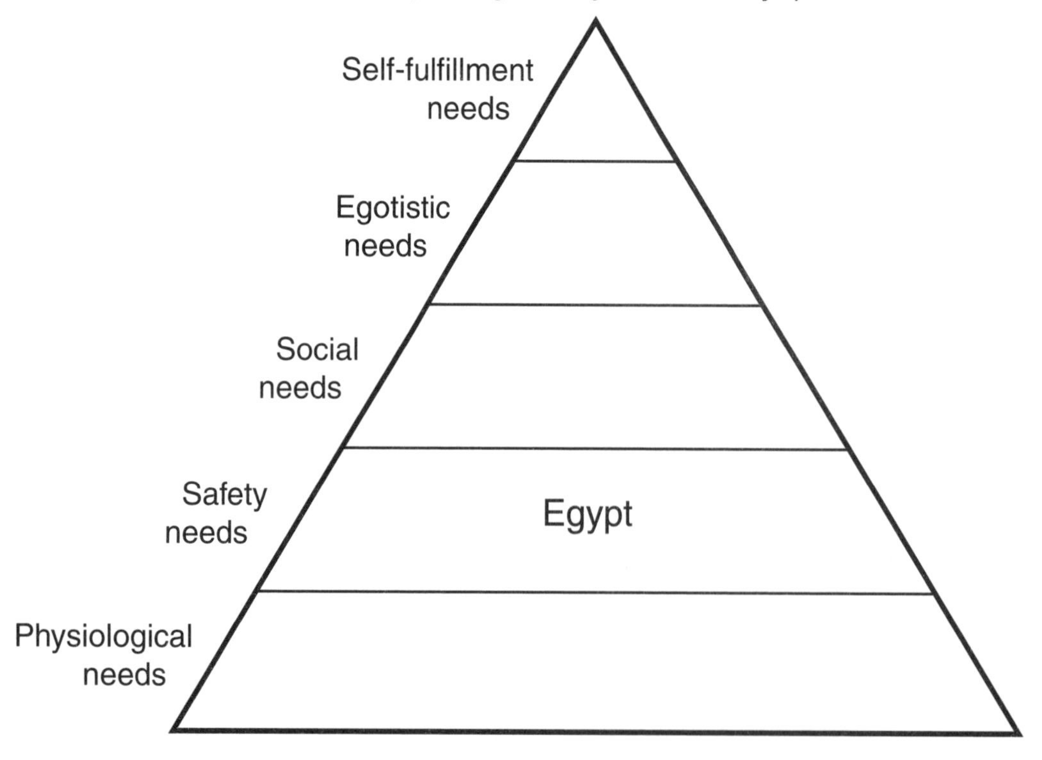

Glossary of Geography Terms

A

Absolute Location: the exact address of a place on the earth's surface; the latitude and longitude coordinates

Agriculture: using lands for keeping and grazing animals or growing crops

Apartheid: policy of separating people of different races in South Africa

Archipelago: group or chain of islands

Arid: dry

B

Bay: part of a sea that is smaller than a gulf and partly surrounded by land

C

Canal: an inland waterway built to carry water for irrigation or transportation

Canyon: deep valley with steep sides and a stream running through it

Civil War: a war between opposing groups of people within the same country

Coast: land along the sea

Colony: area settled or taken over by people from another country

Communism: political system of government in which land, industry, and all property and goods are controlled by the government or state

Conservation: actions taken to protect the natural environment

D

Dam: a piece of land or a structure that is used to hold back the flow of a body of water

Deforestation: loss of areas of forest due to human consumption

Democracy: system of government where the people elect those who govern them

Desert: a dry region where more water is lost due to evaporation than falls as precipitation

Desertification: the spread of a desert due to climate change or mismanagement

Drought: long period of dry weather, possibly leading to a dangerous shortage of water

E

Economy: the way a country produces, uses, and distributes its money, resources, goods, and services

Ecotourism: the practice of touring natural habitats in a manner meant to minimize ecological impact

Equator: imaginary line around the middle of the earth at an equal distance from the North and South Poles—zero degrees latitude

Erosion: the gradual wearing or washing away of rock, land, or buildings by the force of wind and water

Ethnic Group: group of people within a larger community who have different characteristics such as language, religion, or customs

Export: product made in one country and sold to another

Extinct: no longer in existence

F

Famine: extreme shortage of food, possibly resulting in starvation

G

Globe: a spherical model representing Earth as it looks from space

Grid: system of latitude and longitude lines that helps find places on a map

Gross Domestic Product (GDP) per capita: the per-person figure that represents the total value of all goods and services produced in a country in one year

H

Habitat: the specific natural surroundings in which humans, animals, and plants live

Harbor: a deep body of water where ships can anchor

Human Region: an area that can be identified by the characteristics of its people, such as government, religion, or language

I

Import: product that is brought into a country from another country

Interactions (Human-Environment Interactions): the relationship between how humans affect their environment and how the environment affects humans

Irrigation: system for providing a supply of water to crops

Glossary of Geography Terms

J

Jungle: land covered with dense growth of trees, vegetation, vines, etc., typically in tropical regions

K

Key: a reef or low island

L

Latitude: horizontal lines measuring distance north and south of the equator

Legend: tells the meanings of the symbols on a map; also known as a map key

Location: (See Absolute or Relative Location)

Longitude: vertical lines measuring distance east and west of the prime meridian

M

Monarchy: a system of government in which the land is ruled by a king, queen, emperor, or empress

Monsoon: extremely strong seasonal wind accompanying wet, stormy weather

Movement: the study of how people, goods, and ideas have moved, or are currently moving, around the globe

N

Natural Resource: part of the natural environment that humans can use to satisfy their needs and wants

O

Oasis: a fertile place in a desert, due to the presence of water

P

Physical Region: an area defined by its plant life, natural resources, animal life, landforms, or climate

Place: the physical and human characteristics that make a location unique

Political Map: a map that shows boundary lines and borders

Population Density: the number of people per unit of land area

Prime Meridian: imaginary line from which longitude is measured both east and west; zero degrees longitude running through Greenwich, England, from the North Pole to the South Pole

Province: area of land that is part of a country or empire

Q

Quagmire: wet, boggy ground, yielding under the foot

R

Refugee: person forced to flee his or her home country due to war, famine, or political or religious persecution

Region: an area of land that shares one or more common characteristics; can be physical or human characteristics

Relative Location: identifying where a place is in relationship to other places

Rural: in, of, or like the country

S

Scale: shows the measurement of distances and areas on a map

State: large community that is organized under one local or national government

T

Title: tells what the map is supposed to show; the main idea of the map

Topography: physical features

U

Urban: in, of, or like a city

V

Valley: lower land that lies between hills and mountains

W

Wetlands: areas that have wet soils, such as swamps or marshes

X

Y

Z

Zephyr: the west wind; a soft, gentle breeze

Websites and Suggested Activities

WEBSITES:

<www.commoncore.org> - CCSS website
<www.curriculum21.com> - resources for teachers
<www.gapminder.org> - visual tools for students to display data
<www.googleearth.com> - visual depictions of most areas of the world
<www.nationalgeographic.com> - National Geography Standards and Skills
<www.nato.int> - NATO's website
<www.socialstudies.org> - National Council for the Social Studies
<www.un.org> - United Nations website
<www.unicef.org> - UNICEF (State of the Children link)
<www.100people.org> - global education toolbox

SUGGESTED ACTIVITIES: (Teachers and students can adapt these activities to available technology and apps.)

1. Name that Term game:

   Using terms found in the glossary and blank index cards, create a "name that term" game.

   A. Make cards for each of the land and water features. On one side of the index card, write the name and definition of the feature. On the reverse side, draw a picture of the feature.

   B. Make cards for the seven continents by drawing their shapes on one side of the card and writing their names on the reverse side.

   C. Make cards for the four oceans by writing a description of the continents they are located near on one side of the card and writing their names on the reverse side.

2. Create a crossword puzzle.

3. Create a multimedia presentation (i.e., podcast, i-Movie™, Prezi™, etc.) on a current event in one of the world's regions, and connect it to one of the five themes of geography.

4. Create your own country, complete with cities and land features, incorporating the five themes of geography.

5. Choose a world country. Create three maps: topographical, political, and climate. Also include a "fact sheet" on your country, including basic information such as imports, exports, religions, type of government, climates, languages, gross domestic product, and so on.

Answer Keys

United States and Canada: Location—Activity (p. 3)

1. Coordinates may vary by atlas used.
   a. 41°N, 87°W
   b. 39°N, 77°W
   c. 34°N, 118°W
   d. 39°N, 105°W
   e. 33°N, 84°W
   f. 45°N, 75°W
   g. 50°N, 97°W
   h. 45°N, 73°W
2.
   a. New York City, New York
   b. Seattle, Washington
   c. Milwaukee, Wisconsin
   d. Miami, Florida
   e. Edmonton, Canada
   f. Toronto, Canada
   g. Halifax, Canada
   h. Whitehorse, Canada

United States and Canada: Place—Geoquest (p. 8)

Teacher check map.

United States and Canada: Interactions—Geoquest (p. 11)

Answers will vary. Possible answers include:

1. Recycling, water purification, forest preserves, emission control
2. Clearing of forests, pesticides, oil spills, pollution
3. Oxygen, water, food, soil to grow crops, recreation
4. Floods, drought, desertification
5. Answers will vary.

United States and Canada: Movement—Activity (p. 13)

1. Spread of people, goods, and ideas
2. People travel and move. French settled in those areas.
3. Answers will vary.
4. Native Americans, Mormons, slaves, Acadians/Cajuns, etc.
5. Lumber, cotton, automobiles, oil, etc.
6. Democracy, language, religion, etc.

United States and Canada: Regions—Geoquest (p. 17)

Teacher check map.

Central and South America: Location—Activity (p. 19)

Teacher check map.

Central and South America: Place—Questions (p. 22)

1. Tropical climate; located on or near the equator; jungles and wetlands; Amazon River in the interior; Andes Mountains along the western coast from north to south; Argentina and Chile at the southernmost tip have a variety of climates
2. Central America is an isthmus connecting the larger landmasses of North and South America. Mexico is the largest Central American country. It has a warm, arid climate. Mountains are in central and western parts.
3. Large cities are Buenos Aires, Argentina; Rio de Janiero, Brazil; and Santiago, Chile; citizens decended from Native Americans and/or their European conquerors; primary language is Spanish (Portuguese in Brazil); Catholicism is major religion; huts built for warmth in the Andes and for air circulation and sun protection in the Amazon
4. Mexico City is one of the largest cities in the world. Citizens decended from Native Americans and/or their European conquerors; primary language is Spanish; Catholicism is major religion
5. Isthmus: a narrow body of land connecting two larger landmasses

Central and South America: Interactions—Questions (p. 25)

1. Most make their living by farming.

2a-f. coffee, cacao, wheat, sugar, bananas, rice

3. It loses important nutrients crops need to grow, and the land becomes unproductive.
4. Answers will vary.
5. It encourages people to preserve the environment so tourists will come spend money in the area.

Central and South America: Interactions—Geoquest (p. 26)
Teacher check map.

Central and South America: Movement—Questions (p. 28)

1. They crossed a narrow bridge of land between what is now Alaska and Siberia.
2. The Inca Empire

3a. Spanish
b. Portuguese
(Accept any of the other languages given.)

4. Atlantic Ocean (Caribbean Sea) and Pacific Ocean
5. It is the first time an Olympics is held in South America. Goods, people, and money are moving in and out of Brazil.

Central and South America: Regions—Activity (p. 31)
Answers will vary with time. Teacher check.

Europe: Location—Activity (p. 34)
Students may have answers like ...

1. Iceland: part of Europe; island nation; not bordered by other countries, but surrounded by the Atlantic Ocean and the Arctic Ocean
2. Austria: in Europe; bordered by Germany, Switzerland, Italy, Hungary, Slovenia, Czech Republic, Slovenia
3. Portugal: in Europe; bordered by Spain and the Atlantic Ocean
4. Germany: in Europe; bordered by the Netherlands, Belgium, Poland, France, Luxembourg, and Austria
5. Italy: in Europe; boot-shaped; surrounded by France, Switzerland, Austria, Slovenia, the Adriatic Sea, the Ionian Sea, the Tyrrhenian Sea, and the Mediterranean Sea
6. Finland: in northern Europe; bordered by Norway, Sweden, Russia, and the Baltic Sea
7. Sweden: in northern Europe; bordered by Finland, Norway, the Baltic Sea, and the North Sea
8. Macedonia: in southeast Europe; part of the Balkan Peninsula; bordered by Albania, Greece, Bulgaria, Serbia, and Kosovo

9. 49°N, 2°E 10. 47°N, 8°E
11. 40°N, 4°W 12. 47°N, 19°E
13. 53°N, 21°E 14. 51°N, 4°E

Europe: Location—Geoquest (p. 35)
Teacher check map.

Europe: Interactions—Geoquest (p. 41)

1. Hermann's tortoise (Greece)
Why endangered: Road traffic, and people are collecting it for the pet trade
Interactions: Increased road construction and pet trade
2. Pyrenean Desman (France)
Why endangered: Polluted streams
Interactions: Humans polluting; lake and river cleanups; factory environmental laws
3. Pond bat (the Netherlands)
Why endangered: Chemicals used to protect timber have threatened its survival.
Interactions: Humans are using a chemical to protect trees, but in turn are endangering the bat habitat.

4-6. Answers will vary.

Europe: Regions—Activity (p. 46)

1. It shows Germany's major land regions; its topography. It is important for agriculture, industry, trade, and settlement.
2. It determines where people settle and how they use the land.
3. Most agriculture would be in the North German Plain.
4. It shows England's population density. It shows why people settle where they do (population geography). It helps urban planners plan for growth, schools, roads, etc.
5. People may choose where to live based on this; opportunities, access, jobs, near water, schools, crime rate, etc.
6. Most of Great Britain is in the 647 to 1,164 persons per square mile range (250 to 449 per square kilometer).

Russia: Location—Questions (p. 49)

1. Monarchy
2. Europe and Asia

3a. 55°N, 38°E b. 59°N, 30°E
c. 43°N, 132°E d. 48°N, 42°E
e. 54°N, 69°E f. 69°N, 33°E
g. 47°N, 39°E h. 55°N, 83°E

4. South of the Arctic Ocean, west of the North Pacific Ocean, north of China and India, east of Germany/Europe

5. Communism: a political system of government where land, industry, and property are controlled by the state
6. Monarchy: a system of government in which the land is ruled by a king or queen, emperor or empress

Russia: Place—Questions (p. 52)

1. Rolling farmlands; temperate areas in the south near the Black and Caspian Seas
2. Barren tundra; borders Arctic Ocean
3. serf, slave population until late 1800s; hard-working; small villages; Russian language; Russian Orthodox Christianity; large, over-crowded cities; starting a democratic society
4. The people were controlled by the feudal princes, lords, and monarchs.
5. Democracy: a system of government in which the people elect those who govern them

Russia: Place—Geoquest (p. 53)

Teacher check map.

Russia: Interactions—Questions (p. 55)

1. They have stonger pollution controls.
2. Pollution, nuclear disasters
3. Most people farm; fishing provides a living; people vacation in the warmer areas to the south.
4. There are cold, frigid winters; much of the land is barren and unfarmable.

Russia: Interactions—Geoquest (p. 56)

Answers for Your City will vary.

	Moscow, Russia
Summer Climate:	Avg. 60–70°F
Winter Climate:	Avg. 10–20°F
Physical Environ.:	Varies
Housing:	Houses, apartments
Foods:	Varies
Clothing:	Varies
Transportation:	Automobile, train, airplane
Crops grown:	Wheat, livestock
Industries/Products:	Textiles, automobiles, electronics, artwork

Russia: Movement—Questions (p. 58)

1. It helped spread the word of communism and the overthrow of the monarchy.
2. It helped spread the word of democracy and the overthrow of communism. The Russian people became aware of what others in the world enjoyed.
3. Radio, television, and computers brought Western ideas and culture into the Soviet Union.
4. They wanted to control the people so they wouldn't revolt.
5. The military. They wanted to protect the government from revolution or invasion.

Russia: Movement—Geoquest (p. 59)

1a-c. Chemical products, military equipment, textiles, cars

2a-c. Natural gas, coal, oil

3. 142,470,272 (July 2014 est.)
4. $2.553 trillion (GDP 2013 est.)
5. $16.72 trillion (GDP 2013 est.)
   The United States produces much more than Russia.

Answers will change with time.

Russia: Regions—Activity (p. 61)

	U.S.	**Russia**	**Similar/ Different**
Religion:	Many	Russian Orthodox/ Others	Different
Language:	English	Russian	Different
Govt.:	Democracy	Democracy	Same
Climate:	Moderate	Cold	Different
Landforms:	Much farmland	Much barren land	Different
Clothing:	Varies	Dress for the cold	Different

The rest of the answers will vary.

Russia: Regions—Geoquest (p. 62)

Teacher check map.

Southwest Asia and North Africa: Location—Activity (p. 64)

1. Jordan: located in the Middle East; bordered by Israel, Syria, Iraq, and Saudi Arabia
2. Tunisia: located in northern Africa; bordered by Libya and Algeria and the Mediterranean Sea to the north
3. United Arab Emirates: located on the Arabian peninsula; in the Middle East; bordered by Saudi Arabia, Oman, Qatar, and the Persian Gulf

4. Yemen: located on the Arabian peninsula; bordered by Oman and Saudia Arabia; the Red Sea is to the west, and the Indian Ocean is to the south
5. Libya: located in northern Africa; bordered by Egypt to the east, Tunisia and Algeria to the west, Niger, Chad, and Sudan to the south, and the Mediterranean Sea to the north
6. 37°N, 3°E
7. 41°N, 29°E
8. 30°N, 31°E
9. 16°N, 33°E
10. 33°N, 44°E
11. 34°N, 8°W
12. 32°N, 35°E

Southwest Asia and North Africa: Interactions—Activity (p. 70)

Answers will vary according to resources used and student opinion.

Southwest Asia and North Africa: Movement—Activity (p. 73)

1. 1869
2. It is located between the main part of Egypt and the Sinai Peninsula. It connects the Gulf of Suez and the Red Sea in the south to the Mediterranean Sea in the north. It is important because otherwise ships would have to sail around the southern tip of Africa to get from the Indian Ocean to the Mediterranean.
3. Answers will vary, but might include moving ships, troops, tourists, and goods.
4. It has made trade routes shorter and more accessible. Turn-around rates are quicker.
5. War broke out between Egypt and Israel. Arab-Israeli conflicts have led to the security of the canal being threatened.

Southwest Asia and North Africa: Regions—Activity (p. 76)

1. Tiny marine plants and animals decayed, and pressure applied by layers of sediments caused the material to change to petroleum.
2. It has been used as a bargaining chip. Countries who have oil have more trade and wealth, so other countries want to control the oil supplies.
3. Kuwait has cut off oil shipments to the United States when it did not agree with U.S. policies.
4. It is the body of water closest to some of the oil-producing nations, so ships carrying oil sail through the Persian Gulf.
5. It is the Organization of Petroleum Exporting Countries. It sets the price of petroleum.
6. Kuwait is a vital source of oil. The United States did not want the Iraqis to control that oil.
7. Answers will vary.

Africa, South of the Sahara: Location—Activity (p. 79)

1. It forms part of the northern border of Tanzania, and it borders Kenya and Uganda.
2. They are along the southern edge of South Africa.
3. It is an island nation off the eastern coast of southern Africa.
4. It is situated between Malawi, Mozambique, and Tanzania.
5. The equator cuts through the southern part of Somalia. It is bordered by the Indian Ocean to the east and the Gulf of Aden to the north.
6. Antananarivo, Madagascar
7. Cape Town, South Africa
8. Luanda, Angola
9. 34°S, 18°E
10. 16°S, 28°E

Africa, South of the Sahara: Place—Activity (p. 82)

1. Ethiopia
2. Tanzania
3. Dem. Rep. of the Congo (Zaire)
4. Angola

Africa, South of the Sahara: Place—Geoquest (p. 83)

1. Visitors were more free than the blacks living in South Africa. Even though black South Africans were citizens, they had no rights.
2. It caused much conflict, death, and tension between whites and blacks. Other countries placed economic sanctions on South Africa until apartheid ended.
3. The white minority had all the power and controlled the government and economy. They feared what the black majority would do if they had any power.
4. The United States was against apartheid and imposed economic sanctions on South Africa.
5. It was a racial policy of separating the races. The white minority had all the control, and the black majority did not have any power.

6. Nelson Mandela (1918–2013) was a black leader in South Africa who spent 28 years in prison for opposing apartheid and fighting for the right for blacks to vote. In 1994, he became the first black president of South Africa.

Africa, South of the Sahara: Movement—Activity (p. 88)

1–10. Answers will vary.
11. The United States trades with African nations, both exporting and importing.
12. Answers will vary, but may include the United States, South Africa, England, France, and Germany.

Africa, South of the Sahara: Regions—Activity (p. 91)

Answers will vary. Teacher check.

Asia, Oceania, Australia, and Antarctica: Location—Activity (p. 94)

1a–d. Southwest Asia (Middle East), Indian subcontinent, Southeast Asia, Eastern Asia
2. It was far from Europe, and one had to travel east to get there.
3a. 12°N, 104°E
b. 6°S, 107°E
c. 21°N, 105°E
d. 14°N, 121°E
e. 40°N, 117°E
f. 28°N, 77°E
4a. 8,565 miles
b. 9,280 miles
c. 6,314 miles
d. 7,618 miles

Asia, Oceania, Australia, and Antarctica: Place—Questions (p. 97)

1a. Tropical jungles, wetlands, large rivers, desert, mountains
b. Dry and hot with seasonal monsoons; mountains, rivers
c. Farmable islands, volcanic islands, mountains
d. Tropical coastal regions, large inland deserts, farming and grazing land, mountains
e. Tropical islands, jungles, mountains, volcanoes
2a. Ruled by an emperor for centuries, now a communist government
b. Buddhist religion began there, but the majority of people are Hindus
c. Ruled by an emperor for centuries; now a democratic government with the emperor as a figurehead
d. Former British colony; federal constitutional monarchy; English-speaking
e. Have a democratic government; once ruled by Spain and then the United States
3a. Monsoon: period of rainy, stormy weather
b. Monarchy: a system of government ruled by a king, queen, emperor, or empress

Asia, Oceania, Australia, and Antarctica: Interactions—Activity (p. 100)

Answers will vary. Teacher check.

Asia, Oceania, Australia, and Antarctica: Movement—Activity (p. 103)

1. Movement: the study of how people, goods, and ideas move around the globe; Examples may vary.
2. He was an early explorer from Italy who traveled to the Far East. He opened up trade routes between the East and West.
3. Japan became more technological and began exporting its products around the globe.
4. It allows for the easier and quicker spread of news, ideas, goods, and people. Answers will vary.
5. It was built to keep out foreign invaders.
6. Prisoners from England were the first Europeans to live in Australia. They were selected as a punishment.
7. The Norwegian Roald Amundsen was the first to reach the South Pole. Because he reached the South Pole, more people followed, and ultimately, scientific research bases were opened.

Asia, Oceania, Australia, and Antarctica: Regions—Activity (p. 106)

Answers will vary with time. Teacher check.